MW00764210

PETROCHEMICAL TECHNOLOGY

PETROCHEMICAL TECHNOLOGY

An Overview for Decision Makers in the International Petrochemical Industry

H.L. LIST

President, List Associates, Inc.
Consulting Chemical Engineers
Professor Emeritus of Chemical Engineering
City University of New York

PRENTICE-HALL
Englewood Cliffs, New Jersey 07632

Library of Congress Cataloging-in-Publication Data

List, H. L. (Harvey L.), (date)
Petrochemical technology.

Includes bibliographies and index.
1. Petroleum chemicals industry. 2. Petroleum chemicals industry—Technological innovations.
3. Petroleum chemicals. I. Title.
HD9579.C32L44 1986 661'.804 85–19225
ISBN 0–13–661992–4

Editorial/production supervision and
interior design: David Ershun/Nancy Menges
Cover design: Wanda Lubelska
Manufacturing buyer: Rhett Conklin

Printed in the United States of America

10 9 8 7 6 5 4 3 2 1

ISBN 0-13-661992-4 025

Prentice-Hall International (UK) Limited, *London*
Prentice-Hall of Australia Pty. Limited, *Sydney*
Prentice-Hall Canada Inc., *Toronto*
Prentice-Hall Hispanoamericana, S.A., *Mexico*
Prentice-Hall of India Private Limited, *New Delhi*
Prentice-Hall of Japan, Inc., *Tokyo*
Prentice-Hall of Southeast Asia Pte. Ltd., *Singapore*
Editora Prentice-Hall do Brasil, Ltda., *Rio de Janeiro*
Whitehall Books Limited, *Wellington, New Zealand*

This book is dedicated
to the memory of

CLAIRE WINKEEPER LIST

An Original

Contents

Preface

If decision makers in the international petrochemical industry are to compete effectively, they must have a basic understanding of the relationship between major petrochemicals and fossil fuels and the effects of international competition, government actions, and political constraints. The problem is extremely complex. The number of variables is massive, and political and governmental actions are always unpredictble. However, a point of departure for decision making is possible, and the purpose of this text is to provide this basic point.

The worldwide petrochemical industry has been undergoing marked changes in the past few years, and disruptions will unquestionably continue to be the rule for the future. The term *petrochemical* generally includes chemicals and derivatives that are obtained from the processing of crude oil and natural gas. Since the economics of the petrochemical industry are obviously intimately tied to the energy industry, changes in fuel supplies generally affect supplies and prices for petrochemicals shortly thereafter.

About 10 years ago the action of the Organization of Petroleum Exporting Countries resulted in a series of price increases that greatly affected the economics of the petrochemical industry. A perceived decline in reserves of natural gas in the United States further aggravated the situation. U.S. domination of international petrochemical markets is expected to decline during the eighties as huge petrochemical industries develop in places such as Canada, Mexico, and the Middle East, where large supplies of inexpensive feedstocks are available. The major competition is expected to be in primary petrochemicals and commodity products and will affect Japanese and European petrochemical industries as well as those of the United States. To compete, companies will be forced to improve production through new or improved processes, and less efficient plants will continue to be shut down. Processes with high consumption of utilities and raw materials may become economically unattractive.

The material provided in this text has come from published sources and from a distillation of information obtained from my communication with people in the field as well as my 35 years of experience in petrochemical engineering. A substantial amount of the information has been extracted from a bi-weekly newsletter entitled *International Petrochemical Developments*, which I have written since 1980 and is subscribed to by decision makers in industry and government in over 40 countries around the world.

This book contains current technologies, current major producers, approximate economics for major petrochemicals and commodity products, and possible competing technologies. I have resisted the temptation to provide in-depth data. In-depth evaluations, if desired, would involve detailed studies for a particular product. Furthermore, I have attempted to provide a basic understanding of the processes involved without inundating the reader with engineering details.

The producers and estimated capacities shown are from the latest available information. However, this field is very dynamic. Plants are continually shut down, either temporarily or permanently, restarted, sold, or merged, and the literature reports change almost daily. With regard to the economics of the various processes, the data presented are from published economic information as well as from my own studies in the subject. All data have been adjusted to a Gulf Coast location in the United States and to construction in early 1986. Since actual production costs depend on variables such as location, feedstock, utility costs, and cost of capital, these figures are to be considered as only approximations. The basis for the economics is outlined in the appendix.

The book begins with an introductory chapter covering what I believe to be the latest industry trends, and is followed by chapters on the derivatives of the major primary petrochemicals, such as synthesis gas, ethylene and propylene. In those instances where more than one primary petrochemical is needed for production of the derivative, the derivative is included with the feedstock that is required in greater quantity. In addition, there are often several routes to a particular petrochemical, usually involving different feedstocks. In such cases, even though the petrochemical is located in a particular feedstock section, alternative feedstocks are discussed.

At the beginning of each section is a block diagram showing the interrelationship between feedstocks and major derivatives covered in the section. Each section is divided into divisions covering technology, markets, and economics.

My sincere appreciation goes to my many colleagues in the international petrochemical industry for their suggestions and contributions. The staff of the publisher, Prentice-Hall, have been very cooperative in completing this project. Particular thanks go to my son, Ian M. List, who was very gentle with his father during his editing of the manuscript.

H. L. List

International Petrochemical Developments

International Petrochemical Developments, a twice monthly appraisal of products and processes in the international petrochemical industry was founded by Dr. List in 1980. Dr. List has continued to write the newsletter, which is read by executives in over forty countries around the world. These executives are decision makers in government, the petrochemical manufacturing industry, the feedstock supply industry, and financial institutions. Much of the information in this book is based on studies and evaluations done by Dr. List for use in the newsletter.

The newsletter alerts the subscriber to the tremendous changes that are occurring and will continue to occur in the international petrochemical industry. A section called "BRIEFS" alerts the reader to plans for new facilities, shutdown of old facilities, market studies for various petrochemicals, and government actions that will affect the industry. A "DEVELOPMENTS-TO-WATCH" section alerts the subscriber to longer-range research developments that could affect his future operations. From time to time, an entire issue is devoted to a specific petrochemical and outlines current and potentially competing technology and projected markets.

Additional information can be obtained by writing to the publisher:
International Petrochemical Developments
Rickian, Inc.,
56 Roundtree
Piermont, New York 10968
U.S.A.

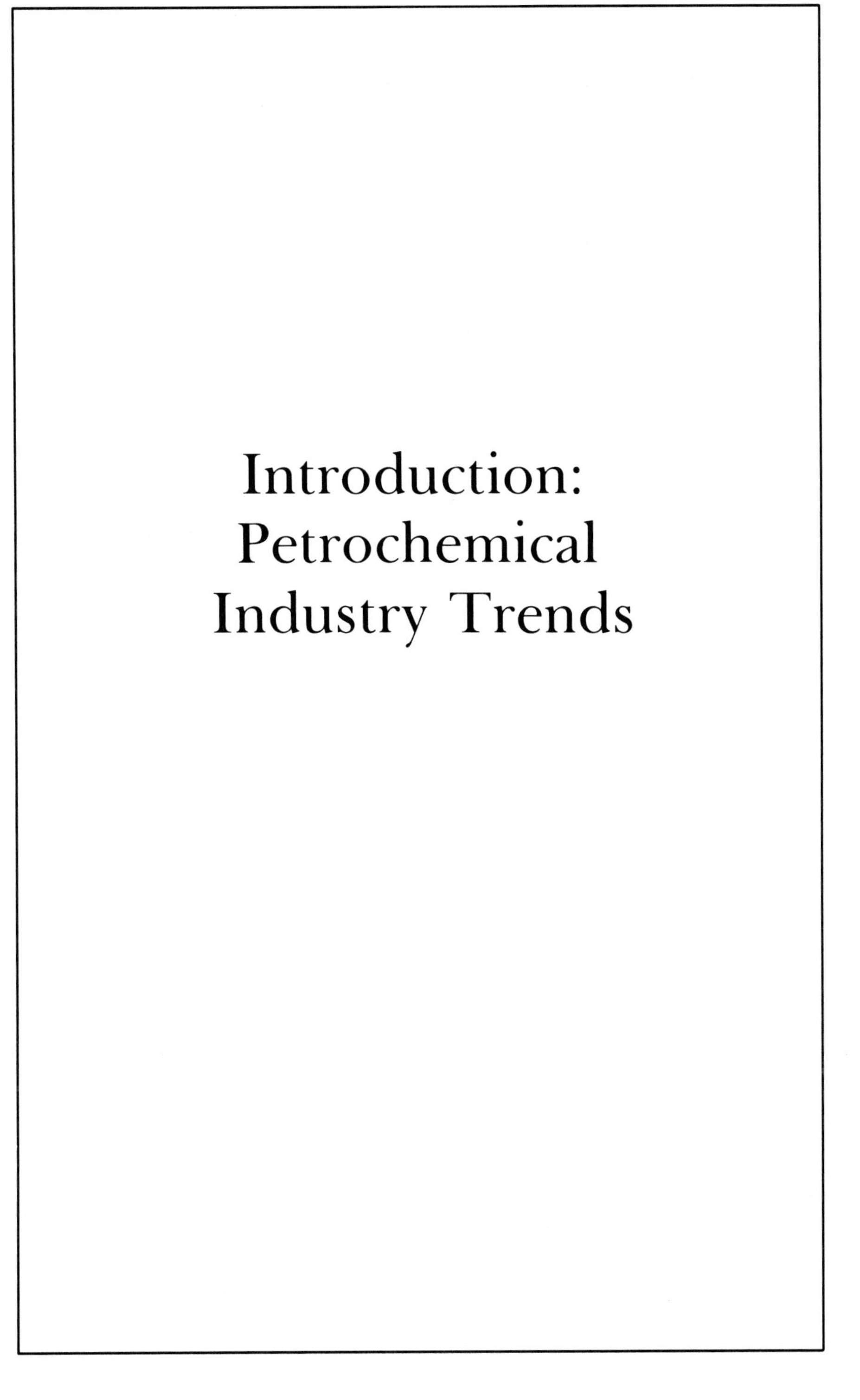

Introduction: Petrochemical Industry Trends

FEEDSTOCK TRENDS

Feedstock use in the petrochemical industry has changed significantly in the last decade, and this trend will continue into the future (1). Worldwide, the olefins-based petrochemical industry developed on the use of surplus hydrocarbon feedstocks from petroleum refinery operations. In the United States, where refinery operations were oriented primarily toward the production of gasoline, natural gas liquids from domestic gas plants have historically been the dominant feedstock for ethylene and still account for about two-thirds of all ethylene production. In Europe, on the other hand, where refining was fuel oriented, straight-run naphtha was a relatively low-priced surplus material, which led to the dominant use of naphtha as the basis for the petrochemical industry.

These days have essentially passed. The petrochemical industry is no longer only an outlet for surplus petroleum cuts. Its feedstocks now play a significant role in the pattern of oil refining and will continue to do so in the future.

The starting point for virtually all petrochemicals involves either (1) steam cracking to produce ethylene and other olefinic products or (2) petroleum reforming and the subsequent extraction of the resulting aromatics. In the steam cracking process, hydrocarbons are thermally cracked with relatively short residence times in the presence of steam using direct-fired tubular furnaces. A wide range of hydrocarbon feedstocks, from ethane to a heavy gas oil, can be used in the steam cracking process. The major product is ethylene, but other products are manufactured, depending on the feedstock used. The three most common feedstocks are natural gas liquids (ethane, propane, butane); naphtha; and heavy gas oil, generally in the 650 to 1000°F boiling range. Typical yields of various products from these feedstocks are shown in the following table.

	Weight % of Feedstock from		
Product	Natural Gas Liquids	Naphtha	Gas Oil
Ethylene	60	30	21
Propylene	10	16	14
Butadiene	2	5	5
Butenes	1	5	6
Steam-cracked naphtha	2	16	19
Fuel products	25	30	35

As can be seen from the table, the major products are olefins. The steam-cracked naphtha is generally high in aromatics. Fuel products include fuel gas and fuel oil.

The cost for the common feedstocks varies considerably, and the general volatility of prices for petroleum products also applies to these feedstocks. They have many uses, and steam cracker feed is seldom the primary use for each product.

In the United States the prices for propane and ethane have generally been lower than the price for either gas oil or naphtha, and the only reason why gas oil and naphtha cracking facilities could compete was the high by-product values for the larger amounts of propylene and butadiene produced. Ethane prices in the United States have been governed by natural gas prices, which are scheduled to be deregulated in 1985. Up to this time the economics have favored the use of natural gas liquids rather than naphtha or gas oil, and this situation is not likely to change, at least for the short term, even after deregulation. Although the complex factors influencing the economics of steam cracker feedstocks make it impossible to predict with certainty the optimum feedstock in the future, traditional feedstocks are expected to be partially displaced some time in the eighties by the increasing use of naphtha and middle distillates. This pattern could continue past the end of the century.

Domestic ethane and LPG supplies will probably not keep pace with increasing ethylene demand and will initially be supplemented by imports of LPG as well as by the use of heavier liquids. Naphtha will probably become a more significant ethylene feedstock as the industry moves further into the eighties. Naphtha, as well as other refinery liquids, could very likely become even more attractive with the phase out of U.S. price controls on natural gas.

Predicted trends in feedstocks for ethylene capacity in the United States are shown in Figure I-1. The percentages of each feedstock are projected in Figure I-2. By the end of the decade, the use of ethane, propane, and butane as feedstocks is expected to fall below 50 percent, with naphtha making up most of the difference. The use of natural gas liquids has required production of

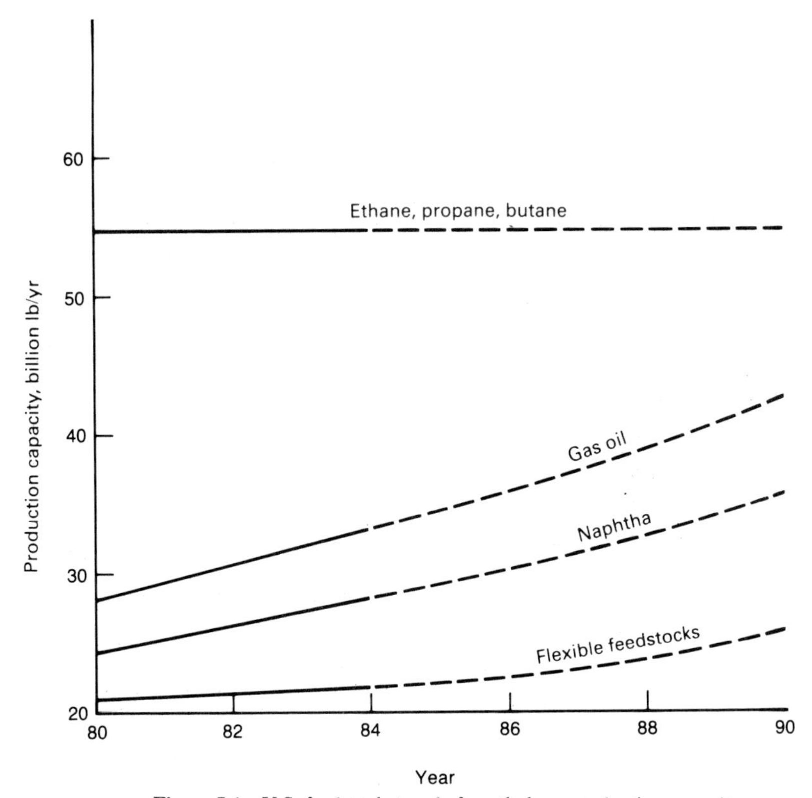

Figure I.1 U.S. feedstock trends for ethylene production capacity

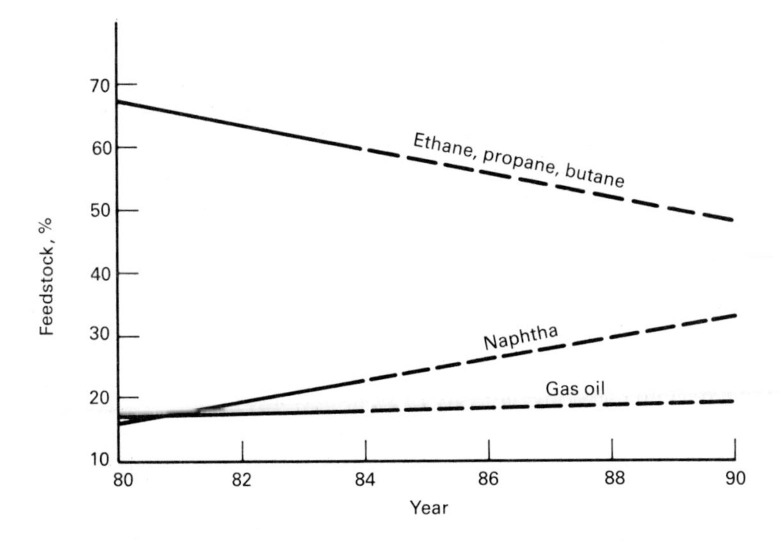

Figure I.2 Feedstock percentages for U.S. ethylene production

propylene primarily from oil refineries and imports of butadiene derived from Western Europe's olefin operations.

In the early stages of the petrochemical industry, demand for feedstocks was very limited relative to the total energy pool, and this imbalance resulted in extremely low prices. However, since the early seventies, the situation has changed markedly as prices for oil and derivatives increased. Although the quantity of naphtha used as a feedstock is much lower than the quantity used as gasoline, there is a complex relationship between the two uses. For example, aromatics are produced in refinery reforming units and from the pyrolysis stream of steam crackers making olefins from heavier liquids. However, aromatics such as toluene and xylenes improve gasoline octane ratings. Predicting what will happen with aromatics is difficult because of the uncertainty in estimating gasoline demand in the future. However, one trend can be projected: The increasing use of unleaded gasoline will undoubtedly contribute to a tightening of aromatics supplies. Although aromatics constitute about 40 percent of olefin production, they are still primarily by-products. About half of the aromatics are produced as a by-product of steam cracking and half as a by-product of the reforming of naphtha to make it suitable for use as gasoline.

A worldwide trend is expected in the use of light-hydrocarbon raw materials—ethane, propane, and butane—to produce ethylene. Substantial quantities of these hydrocarbons are becoming available from areas such as Canada, Mexico, and the Middle East. This trend is expected to be phased in gradually. For the short term the naphtha-based petrochemical industry will probably dominate in Europe. The proportion of ethylene produced from naphtha will decrease gradually, but significantly, as a result of increases in both light-hydrocarbon and gas oil pyrolysis. The decrease will be gradual because the majority of olefin plants are designed specifically for a particular feedstock, with the exception of a few furnaces that can operate on another feedstock. Unless the facility is specifically designed to use multiple feedstocks, producers cannot arbitrarily switch feedstocks from one end of the hydrocarbon spectrum to the other as the relative price structure changes. The only way to ultimately stabilize turbulent feedstock prices is to design the cracking units with the capability to crack different feedstocks. This design is being implemented, but it increases the capital cost of the facility. During the eighties, petrochemical producers will have to be increasingly flexible in their choice of feedstocks, since the variability of prices will greatly affect the cost of production of the petrochemical.

Although the petrochemical industry was developed primarily on the basis of natural gas in the United States and naphtha in Western Europe and Japan, feedstock flexibility is more and more becoming the standard practice. In the United States the trend toward increased use of naphtha and middle distillates will result in closer integration between oil refiners and basic petrochemical producers and will furthermore result in a significant shift in the spectrum of petrochemicals produced. Western Europe, on the other hand, will probably move gradually from a naphtha base toward North Sea deposits of natural gas liquids and perhaps LPG as well as from imports from the Middle East and Africa. Figure I-3 indicates projected feedstock percentages for ethylene production in Western Europe. A drop in naphtha use to about 70 percent of ethylene feedstock is expected by the end of the eighties. The percentage for both light hydrocarbons and gas oil will rise to make up the difference.

Similar changes are predicted for Japan, as shown in Figure I-4. In Japan changes in the negligible use of gas oil as feedstock are not anticipated. As a result of the preceding trends, an increase in the ethylene price ratio between the United States and Japan and Europe is expected. In 1981 the ratio of U.S. ethylene price to European price was about 0.80 and to Japanese price about 0.73. By 1990 both ratios will probably be close to 0.9. For comparison, the price ratios for propylene during the same period are expected to remain fairly constant.

The main sources of energy and feedstocks for the worldwide petrochemical industry will continue to be oil and natural gas through the eighties and probably until the end of the century. At the current level of oil prices, oil- and gas-based feedstocks are still the least expensive among all alternative sources. Over the long term, synthesis gas from coal will provide an alternative, but many authorities do not expect this source to become significant before the end of the

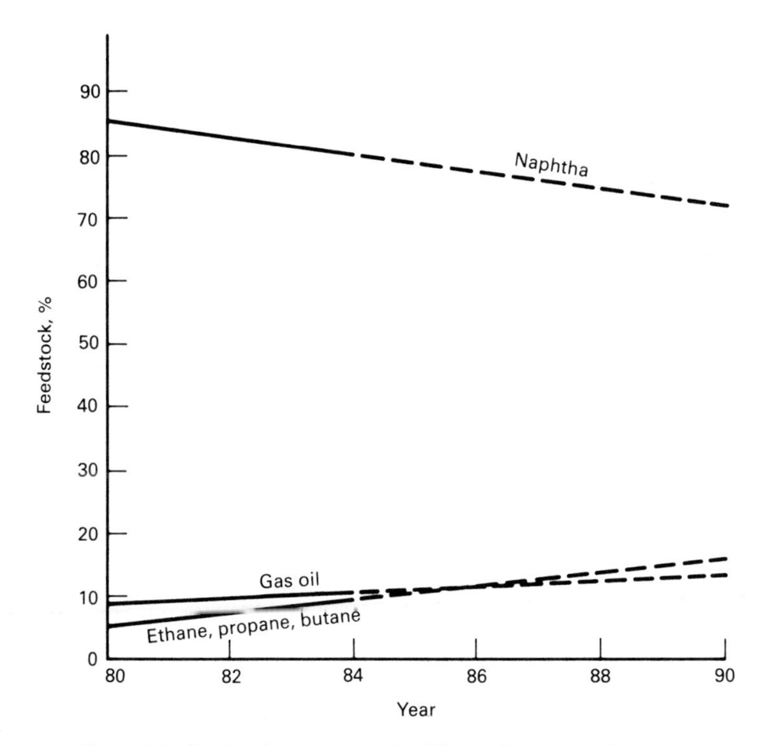

Figure I.3 Feedstock percentages for Western European ethylene production

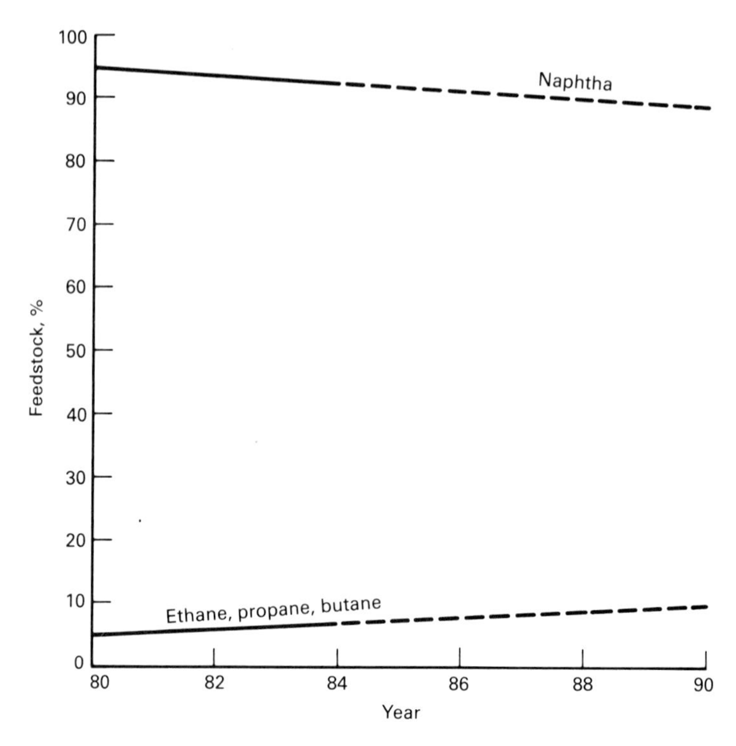

Figure I.4 Feedstock percentages for Japanese ethylene production (Note: Use of gas oil as feedstock is negligible.)

century. Other long-term promising alternatives are the use of methanol and ethanol from biomass as fuel or as petrochemical intermediates.

Another reason for the contention that oil and natural gas will continue to be the major source of petrochemicals for many years is the availability of huge hydrocarbon resources in the oil-producing developing countries. Of particular interest is natural gas, which up until recently was essentially wasted. These resources put these countries in a highly advantageous position to produce basic and intermediate petrochemicals. Figure I-5 predicts trends for ethylene feedstocks in the secondary producing countries, excluding the United States, Western Europe, and Japan. Use of light hydrocarbons will grow to 45 percent by the end of the decade. Both naphtha and gas oil will drop—naphtha to 50 percent and gas oil to 5 percent.

The development of LPG as a petrochemical feedstock will be highly dependent on its price relationship with other feedstocks, primarily naphtha. In 1982 over 20 billion pounds were consumed for petrochemicals, mostly in the United States. As discussed previously, the Western European industry is based primarily on naphtha and some gas oil, with a limited capacity for cracking propane and butane. Use of LPG as a petrochemical feedstock in the United

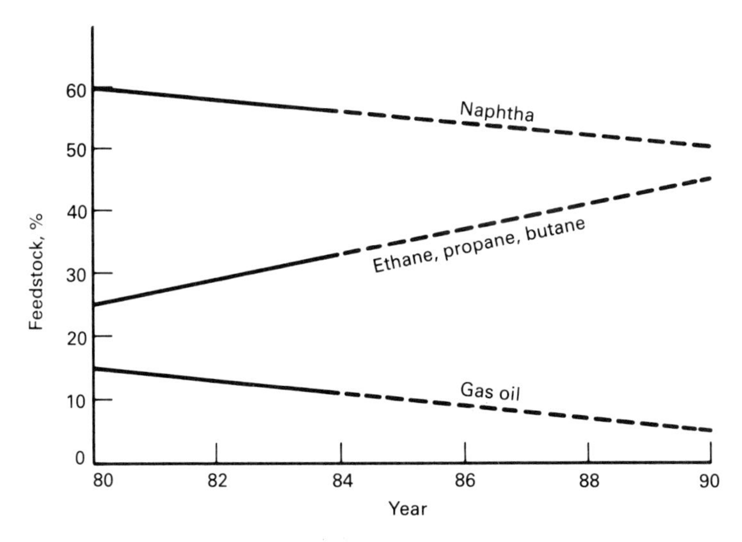

Figure I.5 Feedstock percentages for ethylene production in countries other than the United States, Western Europe, and Japan

States is not projected to grow to any great extent in the eighties. In Western Europe, growth to perhaps 10 billion pounds of LPG per year is expected in the next few years. This level will probably require imports, most likely from the Middle East. As mentioned previously, the potential for this feedstock will depend primarily on its price vis-a-vis naphtha. Some predicted trends in worldwide use of ethane, propane, and butane as ethylene feedstocks are shown in the following table:

	Worldwide Use, billion lb/yr		U.S. Use, billion lb/yr		% World Use by U.S.	
	1980	1985	1980	1985	1980	1985
Ethane	20	34	17	19	85	56
Propane	14	20	13	16	93	80
Butane	7	15	3	4	36	27

U.S. production of ethylene in 1983 was slightly less than 30 billion pounds. With a current installed capacity of approximately 40 billion pounds, this production level would point to a utilization factor of about 75 percent. The somewhat higher demand for ethylene as well as a decrease in installed capacity are responsible for the improvement in the utilization factor since 1982, which was about 65 percent.

As indicated previously, the use of oil-based feedstocks is not expected to be replaced substantially in the eighties. This view is not shared by some authorities, who feel that coal-based chemicals will be competitive in the eighties. Eastman has completed its chemicals-from-coal complex in Tennessee and suggests that coal-derived synthesis gas will be competitive with natural gas or residual oil feedstocks in 1 to 2 years. This competition will probably happen eventually, but in the judgment of many authorities it will not occur until the nineties. The timing will primarily depend on future economic trends and the rates of increase in the costs of crude and natural gas.

When the substitution of coal-based chemicals comes, it will probably come from the use of methanol produced from coal-derived synthesis gas. Processes that will yield important bulk petrochemicals, like olefins and aromatics, directly from syngas will undoubtedly require advanced catalyst systems and will take considerable time before leading to commercial-scale production. Large-scale development of alternative feedstocks from coal, tar sands, shale oil, and so forth will also require huge capital investment, probably requiring government participation. Methanol, perhaps ultimately produced from coal-derived syngas, does show some promise as an olefin cracker feedstock. Of course, it is presently used primarily as a chemical intermediate, motor gasoline constituent, and protein-manufacturing substrate. Methanol can be catalytically cracked, primarily to ethylene and propylene, over specific zeolite catalysts. Mobil's MTG process, a technology for converting methanol to gasoline, could be used with specifically designed catalysts to produce petrochemicals.

Further along the road the use of biotechnology for the production of chemicals will be extensive. Applications will undoubtedly begin with commercial production of small-volume, high-value specialties by the end of the eighties, but large-scale production of olefins and aromatics from this source is not expected until well into the next century. Many technical problems remain.

MARKET SHIFTS

From the standpoint of both demand and supply, the worldwide petrochemical markets will show substantial rearrangements in the balance of the eighties (2). Most of the shift in patterns is expected to occur toward the end of the decade.

At present the petrochemical industry is still concentrated mainly in the United States, Western Europe, and Japan. The United States has been dominant in the international petrochemical markets but is beginning to lose some ground because of the lifting of oil price controls. This trend is expected to continue with the phasing out of gas controls. In addition, massive petrochemical industries are developing in other areas of the world. The result of this development will be diminished exports and increasing imports for the United States as competition is felt from other producers, particularly the hydrocarbon-exporting countries.

Areas of Major Growth

The areas of major petrochemical growth will be concentrated in those parts of the world with ample feedstock supplies. Of these areas Canada, Mexico, the Middle East, and Southeast Asia are expected to be at the forefront of development. Although these regions are not the only areas of substantial development, they are typical and can be considered as examples. All four areas have substantial hydrocarbon reserves. Canada is a developed nation with a substantial population, but the population is still insufficient to absorb the planned petrochemical production. Mexico is a developing nation whose population for the eighties will probably absorb a good share of its petrochemical production. Saudi Arabia, an appropriate example of a Middle East country, is a developing nation with a small population, which cannot absorb the planned petrochemical production. Southeast Asia has some countries with substantial hydrocarbon resources and some countries without resources but with rapidly growing manufacturing facilities. Each of these areas is developing huge petrochemical industries, which will come on stream in the eighties. Because of their limited domestic markets, these countries will have to export and will provide strong competition on world markets.

Canada is pursuing petrochemical development well in excess of the needs of its population. Despite the struggles between the provinces and the federal government, Canada is generally expected to become a substantial worldwide competitor toward the end of the eighties. Government policies have fundamentally altered ownership, production, and costs. The country has a number of potential advantages:

1. Primary feedstock prices will probably be kept low by government control.
2. Extensive quantities of feedstock are available, which include large reserves of natural gas and substantial quantities of ethane and propane.
3. Much of the development is based on joint ventures with multinational companies, which will aid in marketing.

These advantages are expected to outweigh disadvantages such as higher construction costs than in the United States and long distances from ethylene complexes to major markets. The Canadian petrochemical buildup, despite some more recent cutbacks and deferments, has been massive. Most of the projects are being built in western Canada, primarily in the province of Alberta, and many of the facilities will be on stream by the late eighties. At that time the installed ethylene capacity is expected to be over 6 billion pounds per year.

Mexico is, at present, the fourth largest oil producer in the world and has concentrated on basic petrochemicals. From a level of about 4 billion pounds of primarily petrochemicals heavy in ammonia in the early seventies, output is currently close to 20 billion pounds, and a level of well over 30 billion pounds is expected by the end of the decade. Production of basic chemicals is, by law, in the hands of Petroleos Mexicanos (Pemex), the government-owned oil company. The current global oil glut and the resulting decrease in prices have caused a cut in governmental spending on petrochemical facilities. Although this cut has slowed the petrochemical expansion somewhat, the growth continues to be impressive. Demand has essentially kept pace with the expansion, so that petrochemical exports have been relatively small. However, by the end of this decade a substantial amount is expected to be available for export, most likely concentrating in South America, where it will compete strongly with U.S. exports.

The largest concentration of petrochemical expansion in the eighties in the Middle East will be in Saudi Arabia. The government-owned Saudi Basic Industries Corporation (Sabic) is building petrochemical facilities jointly with Celanese and Texas Eastern, Dow Chemical, Shell Oil, Mobil, and a Japanese consortium headed by Mitsubishi. Most of these facilities will be coming on stream in the mid-eighties, and at that time the Middle East will become a substantial producer in the international petrochemical industry. The Middle East has considerable advantages in the production of petrochemicals: primarily inexpensive feedstocks and low-cost, long-term government financing. The plants, as designed, will have to export, and these exports will compete on the international market, particularly in Western Europe. The exports will probably cause overcapacity problems worldwide and are expected to contribute to a lowering of U.S. petrochemical exports.

Another area that bears watching is Southeast Asia. The only countries with substantial domestic hydrocarbons are Indonesia, Malaysia, and Thailand, but manufacturing facilities in other countries in the region are growing substantially. The region is expected to become an increasingly important center of petrochemical manufacturing during the remainder of the eighties. Among the developing areas of the world, Asia has become the leading chemical-exporting region. At present almost half of all chemical exports from developing nations are estimated to be from Asia, with over half of this amount accounted for by the Republic of Korea, Taiwan, Singapore, and Hong Kong.

As worldwide competion begins to intensify, companies are reviewing which of the basic products will best stand the test of the marketplace. Many companies are cutting entire product lines where they do not have a distinct and competitive edge. For example, in the United States Celanese has dropped out of nylon fibers, intermediate chemicals, and producer-textured polyester filament fiber. Diamond Shamrock has dropped out of vinyl chloride and PVC, and DuPont has withdrawn from polyester tire cord. Monsanto has discontinued production of polyester fiber. In addition to dropping out of markets, many companies are diversifying into greater value products, such as specialty chemicals. Furthermore many companies have attempted to bring capacity more into line with projected future demand, which has resulted in abandoned plans and deferred projects.

Growth in Developing Countries

In the eighties the developing countries with hydrocarbon supplies will concentrate on the production of basic and intermediate petrochemicals and commodity plastics. These products are relatively low cost and unsophisticated. This reorientation of the industry will replace the traditional relationship, where the developing countries supply the raw materials and the developed countries provide the manufacturing. By the end of the eighties, the developing countries will supply substantial amounts of basic and intermediate petrochemicals, and the developed countries will increasingly specialize in chemical products requiring a high degree of complex processing and involving capital- and technology-intensive facilities.

The new producers in the petrochemical field, however, are faced with a problem in selling their products on the world market. In today's petrochemical markets a major portion of basic and intermediate chemical products is used captively by the same producer with integrated production processes. Less than half of these products are sold on the open market, and much of these sales are based on long-term contracts. The present share of petrochemical trade of the developing countries is small. These countries are well aware of the dominance of the Western petrochemical market by a relatively small number of companies. Although these countries could enter the merchant market in a small way, they would find it very difficult to export large volumes unless substantial cooperation with the multinational corporations was involved. This cooperation has taken, and will probably continue to take, the form of joint ventures or long-term contracts and would be necessary if massive disruption of the market structure is to be avoided.

Petrochemical Production Trends

The following presentation represents the author's predictions of petrochemical production trends in the developing nations. Although fluctuations due to worldwide economic and political disruptions will undoubtedly occur, the long-term trends should prove to be valid. In all of these projections, two curves are used: (1) the primary producing countries, which include the United States, Canada, Western Europe, and Japan, and (2) the secondary producing countries, which include the USSR, Eastern Europe, and China. The data come from a number of published and unpublished sources. Obviously as one moves further toward the end of the decade, the data are subject to larger errors. However, the trend will continue: the secondary producers will provide a continually growing share of the world's basic and intermediate petrochemicals and the high-volume end products.

Figure I-6 presents the projected production of ethylene, propylene, and butadiene. By the end of this decade, the olefin production for the secondary

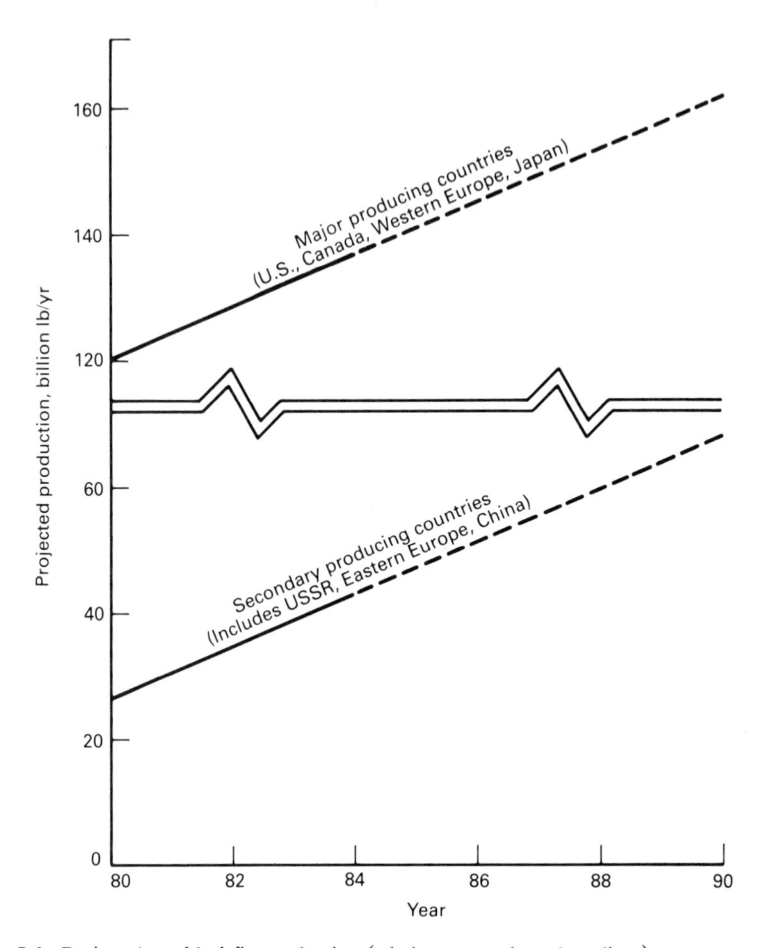

Figure I.6 Projected world olefin production (ethylene, propylene, butadiene)

producers will probably be over 60 billion pounds per year. Figure I-7 shows a similar trend for benzene and xylene. Production of these aromatics will probably reach over 30 billion pounds per year by the end of the decade. Much of the production will likely be used in these same countries to produce intermediate petrochemicals and some high-volume products, such as thermoplastics and synthetic fibers.

Although demand in the developing countries is expected to grow at a higher percentage than demand in the developed nations, capacity will be higher than needed and substantial quantities of material will be available for export to the world markets. Whereas in the past the developing nations have been major importers from the developed regions of such products as polymers, plastic resins, synthetic fibers, and synthetic rubber, the shift in the eighties with the

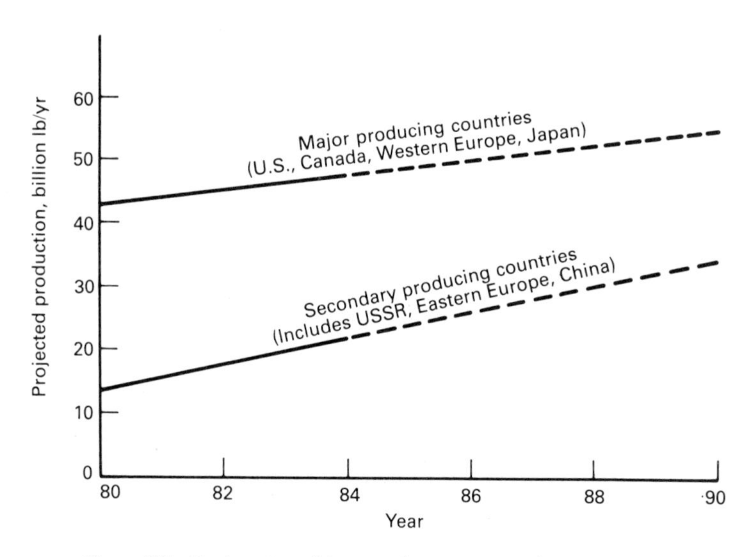

Figure I.7 Projected world aromatics production (benzene, xylenes)

production of these materials will be to export to developed countries and non-oil-producing developing countries.

Figures I-8 and I-9 present the estimated projection trends for major thermoplastics and synthetic fibers. According to predictions, by the end of the decade about 60 billion pounds per year of major thermoplastics will be produced in countries other than the United States, Canada, Western Europe, and Japan. Similarly, these countries will probably produce about 16 billion pounds per year of acrylic, polyamide, and polyester fibers.

Japan, with its lack of oil, will probably be in the weakest position from the competition. Western European producers will face increasing pressure, not only from the Middle East producers but very likely from the increasing production of Eastern European countries. Although one would expect the U.S. petrochemical industry to lose some traditional markets, such as Canada, Mexico, and Latin America, they are probably in a better position to withstand the competition of oil-based developing countries than are other developed regions, since the major U.S. markets are domestic rather than export oriented. However, the U.S. chemical trade surplus is expected to shrink during the remainder of the eighties. Under oil and gas decontrol, U.S. producers will lose an advantage in feedstock and energy costs. It is these rising prices for feedstock and energy in the developed nations that have recently forced, and will continue to force, some petrochemical producers to shut down their plants and import less expensive basic and intermediate petrochemicals from the developing countries. Feedstock and energy costs presently constitute the major share in the cost of production of most petrochemical products. Even with the problem of higher

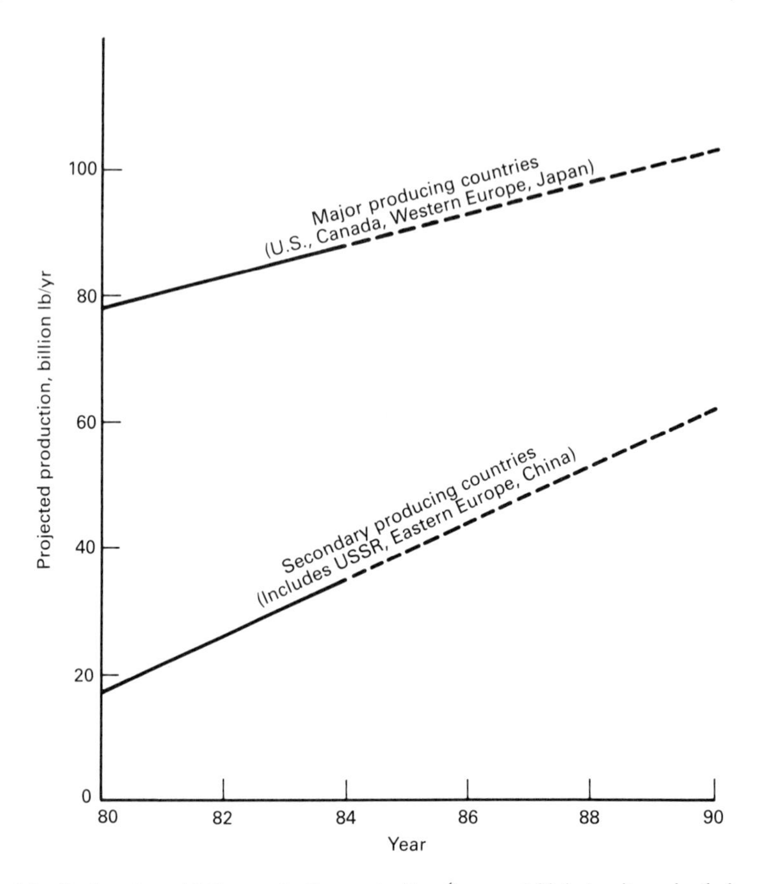

Figure I.8 Projected world thermoplastics production (low- and high-density polyethylene, polypropylene, polyvinyl chloride, and polystyrene)

construction and shipping costs, the developing countries, because of their low-cost raw materials, will have the advantage. They will, however, need access to particular technology and assistance in running the large units demanded by process economics. This assistance has been, and will continue to be, obtained by cooperative agreements with partners from the industrialized nations.

Japan has been a very active participant in new petrochemical projects in developing countries such as Iran and Saudi Arabia. These joint ventures are aimed at ensuring an economic supply of basic petrochemicals in Japan for further processing and, in addition, provide an assured market in these countries for the final products. Note that participants in most of the joint ventures are leading companies in the world petrochemical industry.

The shifts in the industry will also be influenced by the very rapid growth of the economies of some of the developing nations, such as Brazil and South

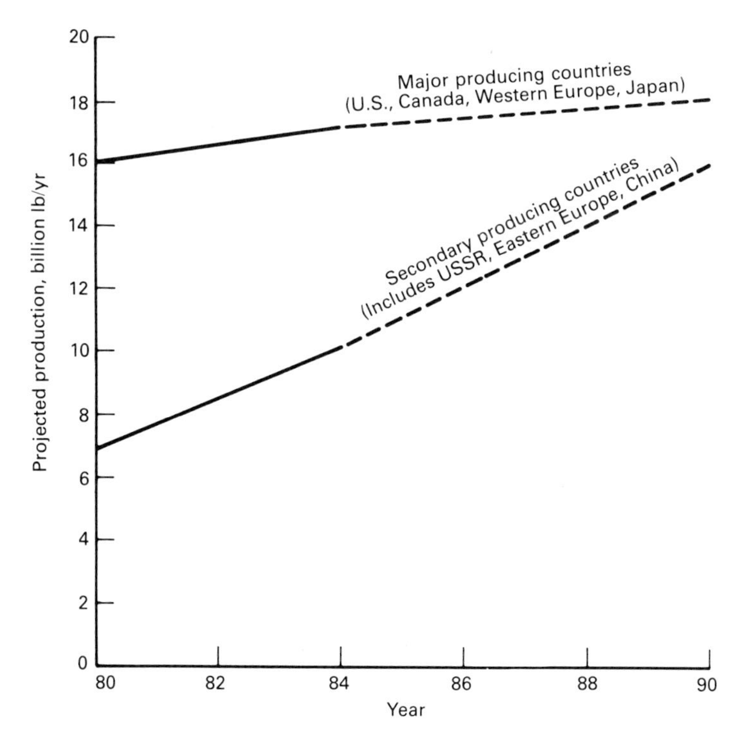

Figure I.9 Projected world synthetic fibers production (acrylic, polyamide, and polyester)

Korea, and major petrochemical expansion plans of the USSR, Eastern Europe, and China.

If the plans of the developing countries with regard to petrochemical production are fully realized, these countries will account for a substantial part of world production, and oversupply is likely to have a tempering effect on prices. The prices will probably rise gradually, but worldwide competition will keep this increase under some control. As the developed nations shut down their unprofitable capacity and as demand and supply are more in balance by the end of the eighties, higher prices will become the rule.

The pattern will be similar to that which has occurred in the international market for ammonia. In the early seventies the United States exported about three times the amount of ammonia it imported. Ten years later the ratio was reversed—the United States imported three times the ammonia it exported. This market reversal has resulted in the shutdown of dozens of U.S. ammonia plants. The situation in Western Europe and Japan is similar. In the eighties the same pattern is expected to prevail for basic and intermediate chemicals as well as for derivatives such as thermoplastics and synthetic fibers.

Figure I-10 summarizes the basic trends in petrochemical production from

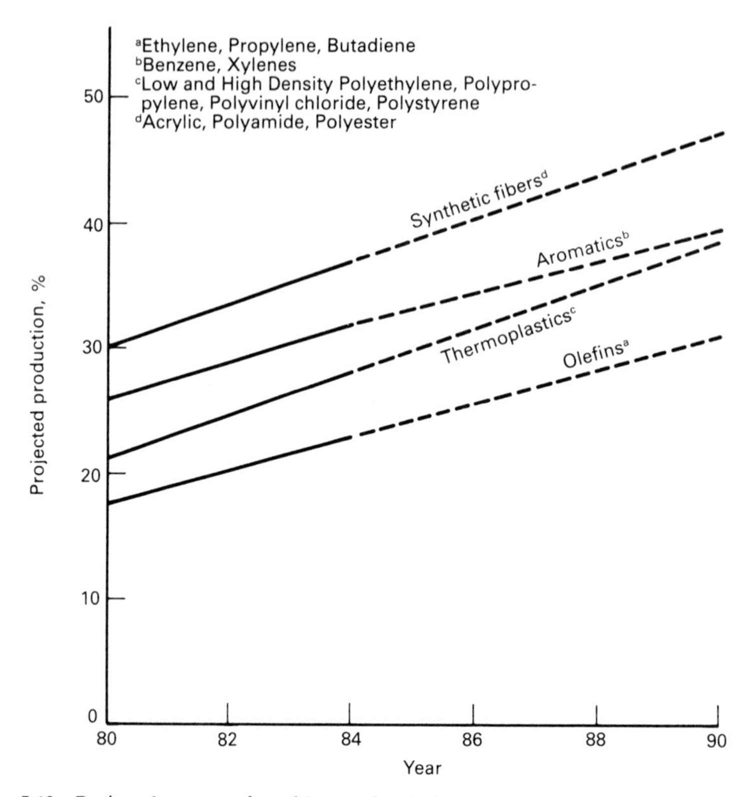

Figure I.10 Projected percent of world petrochemical production in countries other than the United States, Canada, Western Europe, and Japan

the secondary producers. Presented here is the author's judgment regarding the percentage of product that will be attributed to these secondary producers in the near future. These countries will increase their portion of olefin production from about 17 percent in 1980 to about 30 percent by the end of the decade. Aromatics will grow from about 25 percent to about 39 percent. Thermoplastics will grow from about 20 percent to 38 percent, and synthetic fibers from 30 percent to 47 percent. In each category the portion of world production for these products almost doubles.

The established petrochemical producers in the United States, Western Europe, and Japan will have to prepare for the major shift in the market that these trends portend.

REFERENCES

1. List, H. L. "Special Issue—Petrochemical Feedstock Trends," *International Petrochemical Developments,* Vol. 4, No. 16, August 15, 1983.
2. List, H. L. "Special Issue—Petrochemical Market Shifts," *International Petrochemical Developments,* Vol. 5, No. 9, May 1, 1984.

3. "Chemrawn Conference Probes Future of Chemical Feedstocks," *Chemical and Engineering News,* August 20, 1984, pp. 63–67.
4. "Middle East Will Become Major Ethylene Supplier," *Chemical and Engineering News,* July 16, 1984, p. 19.
5. "Soaring Imports Shrink the Surplus in Chemical Trade," *Chemical Week,* June 20, 1984, pp. 14–16.
6. "New Producers Will Force More Rationalization of Capacity," *European Chemical News,* April 16, 1984, pp. 14–16.
7. Uhl, W. C. "Optimism in the Petrochemical Quarter," *Chemical Business,* April 1984, pp. 21–24.
8. "Petrochemical Makers on Comeback Trail," *Chemical Marketing Reporter,* April 2, 1984, pp. 29–44.
9. "International Forecast," *Chemical Week,* January 25, 1984, pp. 64–79.
10. "Petrochemicals '83," *European Chemical News,* December 19, 1983, pp. 3–74.
11. "Forecast 1984," *Chemical Week,* January 4, 1984, pp. 30–41.
12. "World Chemical Outlook," *Chemical and Engineering News,* December 19, 1983, pp. 22–50.
13. "Olefins Momentum Carries Through to '84," *Chemical Marketing Reporter,* January 2, 1984, pp. 27–38.
14. "More European Petrochem. Jolts Seen," *Oil and Gas Journal,* November 21, 1983, pp. 42–44.
15. "Second World-Wide Study on the Petrochemical Industry: Process of Restructuring." Prepared by secretariat of United Nations Industrial Development Organization, May 19, 1981.
16. Ibid., annexes.
17. "How 1985 Adds Up Outside the U.S.," *Chemical Week,* January 23, 1985, pp. 22–32.
18. "ECN European Review," *European Chemical News,* December 1984.
19. "Chemical Profile," *Chemical Marketing Reporter,* January 14, 1985.
20. "Facts and Figures for the Chemical Industry," *Chemical and Engineering News,* June 11, 1984, pp. 32–74.
21. Fairlamb, D. "Saudi Petrochemicals: How Big a Threat," *Dun's Business Month,* February 1985, pp. 66–69.
22. Milmo, S. "Third World Petrochemicals: How Much Market Clout?" *Chemical Business,* March 1985, pp. 11–15.
23. Greek, B. F. "Major Changes Could Confront Basic Petrochemicals This Year," *Chemical and Engineering News,* March 25, 1985, pp. 22–27.
24. "Petrochemicals '85," *Chemical Marketing Reporter,* April 1, 1985, pp. 27–37.
25. "Petrochemicals Supplement," *European Chemical News,* March 1985.

PETROCHEMICAL TECHNOLOGY

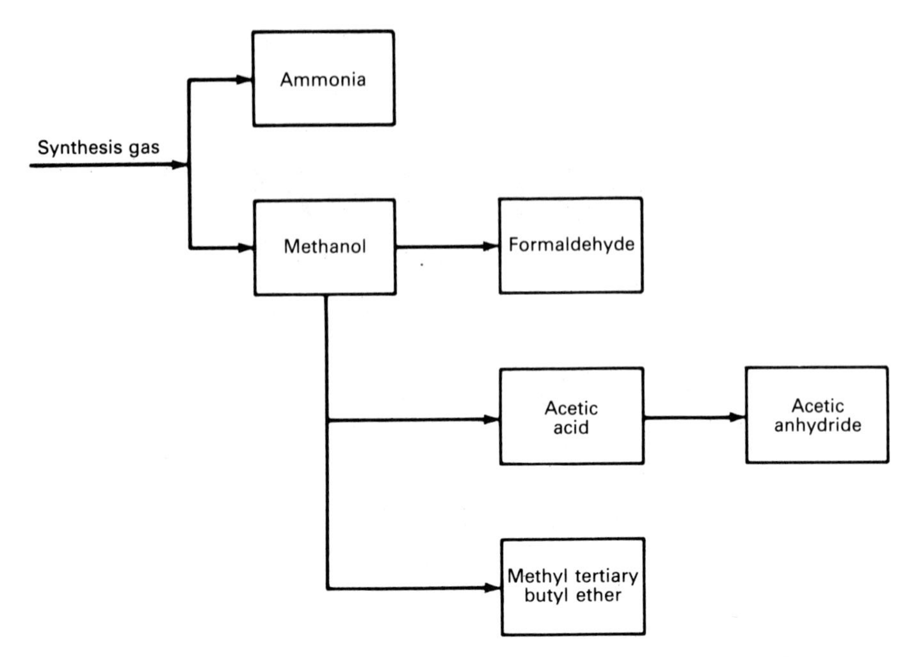

Synthesis gas derivatives

CHAPTER ONE

Synthesis Gas Derivatives

AMMONIA

The conditions affecting the status of the worldwide ammonia business have undergone major changes in the past few years and will continue to change in the future. Changes in the petroleum business have stimulated the development of projects in the hydrocarbon-rich areas of the world. In addition, population and growth projections as well as agricultural development prospects and political considerations impinge on the future of the business.

Technology

The technology in the ammonia industry is expected to evolve relatively slowly. Advanced technologies are not expected to help the traditional producers in the developed countries to compete with the low-cost gas producers. Some high-feedstock producers will scrap some of their plants in favor of those with the highest energy efficiencies. There is a strong possibility that coal-based facilities will be constructed in the United States, Western Europe, and perhaps Japan during the nineties. Therefore the ammonia industry is not expected to enjoy high profits for some years. The control of natural gas prices in the United States has helped domestic producers to compete with producers in Western Europe and Japan, whose feedstock costs have been substantially higher. However, this advantage will disappear as price decontrol for natural gas continues in the United States.

All synthetic ammonia production is currently based on the Haber-Bosch synthesis, which involves the reaction of hydrogen and nitrogen at high pressure over a promoted iron catalyst. The major difference among the various processes is in the method by which the hydrogen-nitrogen gas is produced, and the method used depends primarily on the feedstock. Light hydrocarbons, from methane to naphtha, are generally steam reformed, whereas heavier feedstocks generally use partial oxidation. The technology has advanced primarily because of the introduction of centrifugal compressors and the consequent increase in plant sizes to approximately 1500 metric tons per day. Future developments will probably center on improving the efficiency of the existing processes. The most efficient plants in terms of overall energy consumption are based on natural gas feedstocks.

The most common, and generally the most economical, way to produce synthesis gas is the catalytic reaction with steam at high temperatures. The lighter hydrocarbons are generally suitable for steam reforming, but heavy hydrocarbons tend to form coke on the catalyst. The overall reactions using methane are as follows:

$$CH_4 + H_2O \longrightarrow CO + 3H_2$$

$$CO + H_2O \longrightarrow CO_2 + H_2$$

The major process steps for steam reforming include the following:

1. Desulfurization of the feedstock.
2. Reaction of the feed with steam over a catalyst.
3. Reaction of the reformed gas with air over a catalyst.
4. Conversion of the carbon monoxide in the reformed gas to carbon dioxide by reaction with steam over a catalyst.
5. Removal of the carbon dioxide by absorption.
6. Conversion of the residual carbon oxides to methane over a catalyst.
7. Compression of the gas to synthesis pressure with interstage removal of water.

Figure 1-1 indicates a block flowsheet for the process. The modern steam reforming facility is very energy efficient. Steam is produced wherever possible in the plant, and heat is recovered to preheat boiler feed water. The steam produced is used in the process and to drive turbines.

Partial oxidation processes using fuel oils as feedstock have been developed and licensed widely by Texaco and Shell. They are designed to process whole crudes and heavy residual oils, which have a substantial sulfur and metals content, into a hydrogen-rich synthesis gas for ammonia production. Although this process can use lighter feedstocks, it is not generally considered for such feedstocks. Partial oxidation involves the combustion of a hydrocarbon feedstock in a flame in the presence of less than a stoichiometric amount of oxygen to

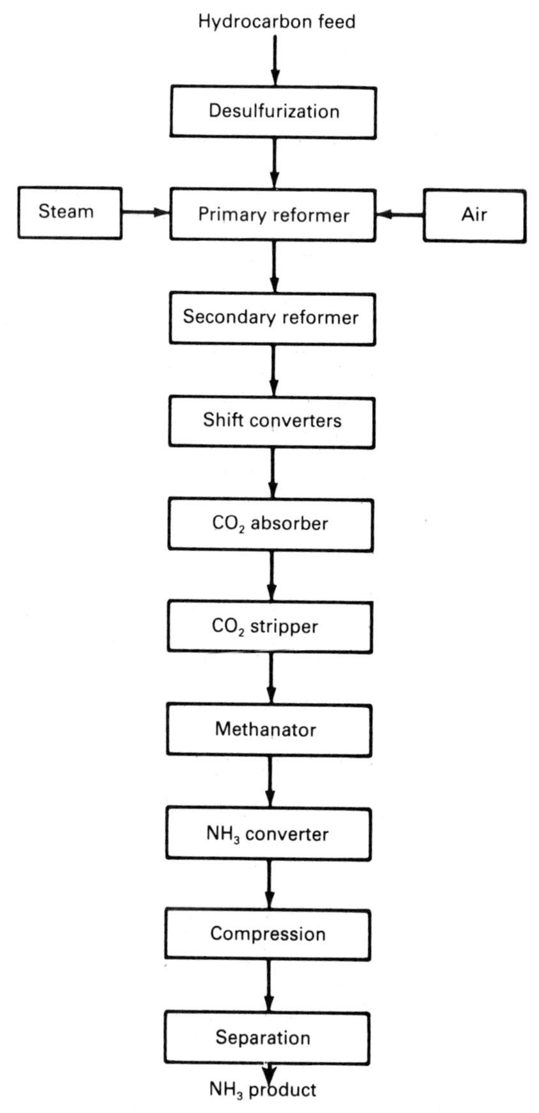

Figure 1.1 Steam reforming process for ammonia

form carbon dioxide and steam, which then reacts with unconverted hydrocarbon feed to produce carbon monoxide and hydrogen. The reaction involves three stages: heating and cracking, reaction, and soaking. Partial oxidation processes are typically operated at pressures between 45 and 90 atmospheres, and the resulting synthesis gas is adjusted to produce a suitable feed for ammonia production. The major difference involves how the hot gases from the gasification reactor are quenched. The quench may be either a direct quench with water or the removal of heat in a waste heat boiler. The direct quench is used for the Texaco technology and the waste heat boiler by Shell, but the difference is gradually disappearing. Figure 1-2 is a block flowsheet for a typical partial oxidation process.

Market

The availability and costs of feedstock are of major importance to the ammonia business. The shifts that are occurring are primarily due to the effects of feedstock, and ammonia production is increasingly concentrated in areas having access to low-cost natural gas. The United States, Western Europe, and Japan are therefore becoming more dependent on imports. High world energy prices favor the gas-rich exporting regions despite higher capital costs and other disadvantages, since the feedstock cost advantages are generally more important in the cost of production.

The market for ammonia is expected to grow steadily because of the large, unsatisfied potential demand in many regions of the world due to population growth and the pressures to grow more food. The demand is expected to continue to show a shift from the developed to the developing countries of the world. Production is expected to increase more rapidly in the gas-rich OPEC and Eastern European regions than elsewhere. Overcapacity is expected to continue for the remainder of this decade on a global basis, but traditional exports from the developed to the developing regions will decrease markedly and in some cases will be reversed. A prime example of this shift is the increasing U.S. imports of ammonia from Mexico. Supply will become increasingly concentrated in the hands of state-controlled enterprises having the political power to set prices for both feedstocks and products at levels below those required by normal commercial considerations.

Many of the countries that have substantial gas have already built large ammonia and urea facilities and are active on the worldwide market. Many, but not all, of these countries have a limited domestic market, and substantial plant construction will require substantial exports. Prices will probably be held down because of a global oversupply. To secure outlets for ammonia, many of the countries will be involved in joint ventures. The countries of Eastern Europe have for many years been exporting solid nitrogen fertilizers, and ammonia exports, primarily from the USSR, are expected to grow rapidly. Of course, the USSR has large natural gas resources and must export to generate hard currency. In addition, they have complete flexibility to vary feedstock and product prices as desired. This flexibility has permitted them to establish a market base in Western Europe and the United States, as well as in other countries.

In the developing nations domestic ammonia capacity will generally be constructed when the demand can justify an economically sized facility. Self-sufficiency is the objective, since in most cases hard currency limitations often restrict the amount of material that can be imported.

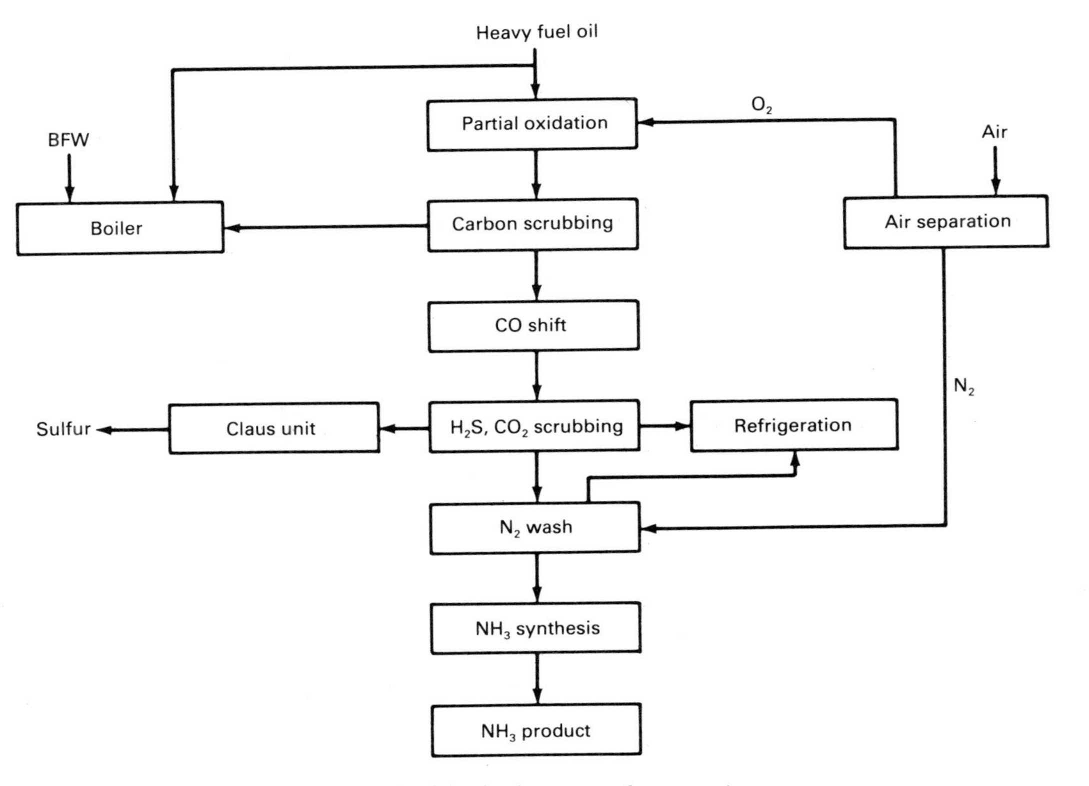

Figure 1.2 Partial oxidation process for ammonia

In the determination of whether a product can compete, all relevant costs must be considered. For example, anhydrous ammonia is shipped worldwide in refrigerated vessels, and the cost of shipping to market is important in the competition between alternative sources of supply.

The major producers of ammonia and their estimated capacities are shown in Table 1-1 for the United States (1), Table 1-2 for Western Europe (2), and Table 1-3 for Japan $(6)_1$.

TABLE 1-1 Major U.S. Producers of Ammonia in the United States

Producer	Location	Estimated Capacity, m MT/yr
Air Products & Chemicals	New Orleans, Louisiana	220
	Pensacola, Florida	90
Allied	Bellevue, Nebraska	162
	Geismar, Lousiana	310
	Hopewell, Virginia	315
American Cyanamid	Avondale, Louisiana	560
Ampro	Donaldsonville, Louisiana	408
Borden	Geismar, Lousiana	370
Cargill	Columbus, Mississippi	60
CF Industries	Donaldsonville, Louisiana	1479
	Terre Haute, Indiana	135
Coastal	Cheyenne, Wyoming	95
Columbia Nitrogen	Augusta, Georgia	490
Cominco American	Borger, Texas	365
Diamond Shamrock	Sunray, Texas	140
Dow Chemical	Freeport, Texas	105
DuPont	Beaumont, Texas	310
	Victoria, Texas	90
Farmland Industries	Dodge City, Kansas	210
	Enid, Oklahoma	860
	Fort Dodge, Iowa	210
	Hastings, Nebraska	145
	Lawrence, Kansas	345
	Pollock, Louisiana	430
Gardinier Big River	Tampa, Florida	125
Georgia Pacific	Plaquemine, Louisiana	170
Getty Oil	Clinton, Iowa	200
International Minerals	Sterlington, Louisiana	430
Jupiter Chemicals	Lake Charles, Louisiana	70
Kaiser	Pryor, Oklahoma	95
Mississippi Chemical	Yazoo City, Mississippi	375
Monsanto	Luling, Louisiana	620
N-Ren	Pryor, Oklahoma	85
	East Dubuque, Illinois	215

TABLE 1-1 Major U.S. Producers of Ammonia in the United States (*continued*)

Producer	Location	Estimated Capacity, m MT/yr
	Carlsbad, New Mexico	60
Oklahoma Nitrogen	Woodward, Oklahoma	430
Olin	Lake Charles, Louisiana	445
Phillips Pacific	Finley, Washington	140
Phillips Petroleum	Beatrice, Nebraska	205
Reichhold Chemicals	St. Helens, Oregon	73
J. R. Simplot	Pocatello, Idaho	91
Chevron	Pascagoula, Mississippi	430
Sohio Chemical	Lima, Ohio	425
TVA	Muscle Shoals, Alabama	67
Terra Chemicals Int'l.	Sergeant Bluffs, Iowa	210
Triad Chemicals	Donaldsonville, Louisiana	315
Tyler	Joplin, Missouri	120
U.S. Steel	Cherokee, Alabama	160
	Geneva, Utah	65
Unocal	Brea, California	236
	Kenai, Alaska	998
The Williams Companies	Blytheville, Arkansas	335
	Donaldsonville, Louisiana	385
	Verdigris, Oklahoma	860
	TOTAL	16,679

TABLE 1-2 Major Western European Producers of Ammonia

Producer	Location	Estimated Capacity, m MT/yr
Chemie Linz	Linz, Austria	643
ASED	Willebroek, Belgium	130
Fison-UCB	Oostende, Belgium	82
SBA Chimie	Vilvoorde, Belgium	140
Soc. Carbochemique	Tertre, Belgium	320
Kemira Oy	Oulu, Finland	230
SARL	Le Grand-Quevilly, France	600
Azote et Prod. Chim.	Toulouse, France	330
COFAZ	Pierrefitte-Nestalas, France	240
Generale des Engrais	Lannemezan, France	100
PEC-RHIN	Ottmarsheim, France	180
Soc. Chim. de la Gr. Par.	Montoir-de-Bretagne, France	100
	Waziers, France	242

TABLE 1-2 Major Western European Producers of Ammonia (*continued*)

Producer	Location	Estimated Capacity, m MT/yr
Societe de L'Ammoniac	Carling, France	330
SARL	Grandpuits, France	300
Soc. Normande de l'Azote	Gonfreville-L'Orcher, France	300
Ammoniak. Brunsbuttel	Brunsbuttel, W. Germany	550
BASF	Ludwigshafen, W. Germany	1120
Chemische Werke Huls	Bottrop, W. Germany	160
	Gelsenkirchen, W. Germany	405
	Herne, W. Germany	250
EC Erdolchemie	Koln, W. Germany	300
Gewerkschaft Victor	Castrop-Rauxel, W. Germany	280
Hoechst	Frankfurt, W. Germany	148
Aeval	Ptolemais, Greece	130
Esso Pappas	Thessaloniki, Greece	115
Nitrigin Eireann Teoranta	Marino Point, Ireland	435
Anic	Gela, Italy	220
	Mont-Sant'Angelo, Italy	360
	Ravenna, Italy	310
Montedison	Ferrara, Italy	470
	Porto-Marghera, Italy	160
	Priolo, Italy	300
	San-Giuseppe-Di-Cairo, Italy	170
Terni Ind. Chimiche	Nera-Montoro, Italy	136
	Porto-Torres, Italy	63
Ammoniak Unie	Rotterdam, Netherlands	310
Esso Chemie	Rozenburg, Netherlands	500
Nererlandse Stik. Maat.	Sluiskil, Netherlands	690
Unie van Kunstmest.	Ijmuiden, Netherlands	280
	Geleen, Netherlands	600
Norsk Hydro	Glomfjord, Norway	120
	Porsgrunn, Norway	370
	Rjukan, Norway	115
Norsk Koksverk	Mo-I-Rana, Norway	60
Petro. e Gas de Port.	Lisbon, Portugal	255
QUIMIGAL	Barreiro, Portugal	73
ENFERSA	Cartegena, Spain	240
	Puertollano, Spain	190
Sociedad Anonime	Elvina, Spain	100
	Malaga, Spain	110
Union Exp. Rio Tinto	Palos-de-la-Frontera, Spain	300
Supra	Koping, Sweden	60
ICI	Billingham, U.K.	1147
	Immingham, U.K.	168
	Severnside, U.K.	295
UFK Fertilizers	Ince, U.K.	280
	TOTAL	16,832

TABLE 1-3 Major Japanese Producers of Ammonia

Producer	Location	Estimated Capacity, m MT/yr	Feedstock
Mitsui Toatsu Chemicals	Chiba	49	Natural gas
	Osaka	500	Butane
Kashima Ammonia	Kashima	320	Waste gas
Showa Denko	Kawasaki	190	Naphtha
Mitsubishi Gas Chemical	Matsuhama	260	Natural gas
Nissan Chemical Ind.	Toyama	160	Naphtha
Mitsubishi Chemical Ind.	Mizushima	90	Waste gas
	Kurosaki	330	Naphtha
Ube Industries	Sakai	310	Naphtha
Ube Ammonia	Ube	310	Naphtha
Sumitomo Chemical	Niihama	270	Naphtha
Nippon Steel Chemicals	Tobata	50	COG
Asahi Chemical Ind.	Mizushima	260	Waste gas
Asahi Glass	Chiba	45	Natural gas
	Kitakyushu	55	Butane
Toyo Soda	Nanyo	55	Crude oil
Tokuyama Soda	Tokuyama	65	Crude oil
	TOTAL	3319	

Economics

The approximate production costs for ammonia from natural gas feedstock are shown in the following table. Almost two-thirds of the cost of production is due to the cost of raw materials, virtually all natural gas, and therefore the natural-gas-rich countries of the world obviously have a substantial cost advantage in ammonia production.

Capacity: 730 mm lb/yr (331 m MT/yr)
Capital cost[a]: BLCC, $89 mm; OSBL, $44 mm; WC, $26 mm

	¢/lb	$/MT	%
Raw materials[b]	6.6	146	62
Utilities	0.5	11	5
Operating costs	0.9	20	8
Overhead costs	2.7	60	25
Cost of production	10.7	237	100
Transfer price	16.2	357	

[a] BLCC: Battery Limits Capital Cost; OSBL: Outside Battery Limits; WC: Working Capital.

[b] Natural gas at $4 per mm BTU

METHANOL

The worldwide methanol business is expected to undergo substantial changes during the next decade—a continuation of the turmoil experienced in the early eighties. In the early eighties a global economic recession affected the growth of the traditional chemical markets, and lowered crude oil prices have slowed the development of new fuel-related markets for methanol. The projected changes are expected to involve all areas of the business, including uses, raw materials and technology, and producers and international trade (1).

Technology

Methanol can be produced from almost any hydrocarbon or carbon material. At present virtually all synthetic methanol is produced by the catalytic reaction of carbon monoxide and hydrogen made either by reforming natural gas or from petroleum-derived heavy-hydrocarbon mixtures. This situation is not expected to change substantially during the next decade. Nonpetroleum sources for methanol are not likely to make significant inroads until the mid-nineties, despite reports of studies for large coal-based methanol plants. In fact, Eastman Kodak in the United States is now producing methanol by coal gasification using Appalachian coal and the Texaco process high-pressure coal gasifier.

A number of technologies are available, but the key is economics. A facility to produce 500 million gallons per year is considered necessary to achieve the lowest cost per unit of production. Such a facility could cost about 2 billion dollars, and there would have to be a substantial price differential between natural gas and coal to justify a project of this size. Despite the deregulation of natural gas in the United States, the economics are not expected to be favorable until at least the mid-nineties.

What is happening, however, is that the methanol facilities are being built in those areas of the world that have an excess of natural gas—places such as Saudi Arabia, Canada, and Mexico. This trend is expected to continue. In many of these regions, natural gas pricing is at the discretion of the government. Methanol can easily be shipped from these sources to world markets. Special ships for cryogenic control, which are required for shipping LNG, are unnecessary for methanol. In some instances barge-mounted methanol plants have been built in areas of established fabrication facilities and towed to remote areas close to the source of the natural gas. For methanol to become a key building block for coal, and therefore for gasoline and chemicals, it must overcome the problem of huge capital costs. Methanol must be able to be made from coal for less than it would cost to make it from natural gas.

Methanol is produced by the reaction of synthesis gas over a promoted catalyst. Practically all recent and planned natural-gas-based methanol projects will use low-pressure synthesis technology, such as that offered by ICI and

Lurgi. The key variable affecting methanol production costs is feedstock costs. Synthesis gas for methanol production can be produced from natural gas, LPG, naphtha, vacuum bottoms, coal, and so forth. For lighter feedstocks through naphtha, the methanol is usually obtained by steam reforming, whereas heavier materials and coal usually require the partial oxidation process.

Steam reforming involves the catalytic reaction of the hydrocarbon feedstock with steam at high temperatures. The overall production of methanol involves desulfurization of feedstock, steam reforming, carbon dioxide addition and syngas compression, methanol synthesis, and methanol purification. The sulfur must be removed from the feedstock because its presence results in the poisoning of the reforming and the methanol synthesis catalysts. In general the sulfur content of the feedstock must be reduced to less than 0.4 ppm by weight before entering the reformer furnace. The relatively low amount of sulfur in natural gas is often removed by adsorption over a zinc oxide adsorbent.

The reforming step uses a nickel catalyst, and the steam reacts with the feedstock at elevated temperatures and pressures to yield a reformer gas containing carbon dioxide, carbon monoxide, hydrogen, and methane. The relative amounts of the components depend on the ratio of steam to hydrocarbon passing over the catalyst and on the temperature and pressure. The overall reaction is endothermic. Steam and natural gas are generally mixed in a ratio of about 2.8 moles of steam per mole of carbon contained in the natural gas feedstock. The reactions are quite complex but can be simplified into the steam-methane reaction and the water-gas shift reaction as follows:

$$CH_4 + H_2O \longrightarrow CO + 3H_2$$

$$CO + H_2O \longrightarrow CO_2 + H_2$$

The necessary heat of reaction is provided by burning fuel in burners, which are located in the furnace and are generally designed to handle natural gas alone or a combination of gases. The heat in the flue gases leaving the furnace is recovered.

When natural gas is the feedstock, the resulting syngas is deficient in carbon, and carbon dioxide is often added to alleviate the deficiency. After the introduction of the carbon dioxide, the syngas is compressed because the methanol synthesis generally operates at higher pressures than steam reforming does.

The formation of methanol involves the reaction between hydrogen and the carbon dioxide and carbon monoxide, and the stoichiometry indicates that the ideal synthesis gas should have a hydrogen content equal to twice the carbon monoxide content plus three times the carbon dioxide content. The reactions are as follows:

$$CO + 2H_2 \longrightarrow CH_3OH$$

$$CO_2 + 3H_2 \longrightarrow CH_3OH + H_2O$$

The methanol converter contains the methanol synthesis catalyst. In the ICI design the methanol synthesis occurs as the gas passes downward through the catalyst. The converter generally contains a number of catalyst beds, and gas is injected at some intermediate points. In the Lurgi process a tubular reactor is generally used, and steam is generated on the shell side as the gases pass downward through the tubes, which are filled with catalyst.

In addition to water, several organic chemicals are produced at the same time as the methanol and are condensed with the methanol. These materials are generally separated from the methanol by distillation. Some of the impurities are ketones, benzene, higher molecular weight paraffins, ethanol and other alcohols, and dimethyl ether. Figure 1-3 indicates a block flow diagram for the process.

Partial oxidation technology for the production of methanol synthesis gas from heavier feedstocks and coal is licensed primarily by Texaco and Shell as well as BASF. The technology involves the following steps: synthesis gas generation and scrubbing, soot recovery and recycling, acid gas removal and shift conversion, and carbon dioxide removal. The reactors are generally refractory lined, vertical pressure vessels. About 2 to 3 percent by weight of the total

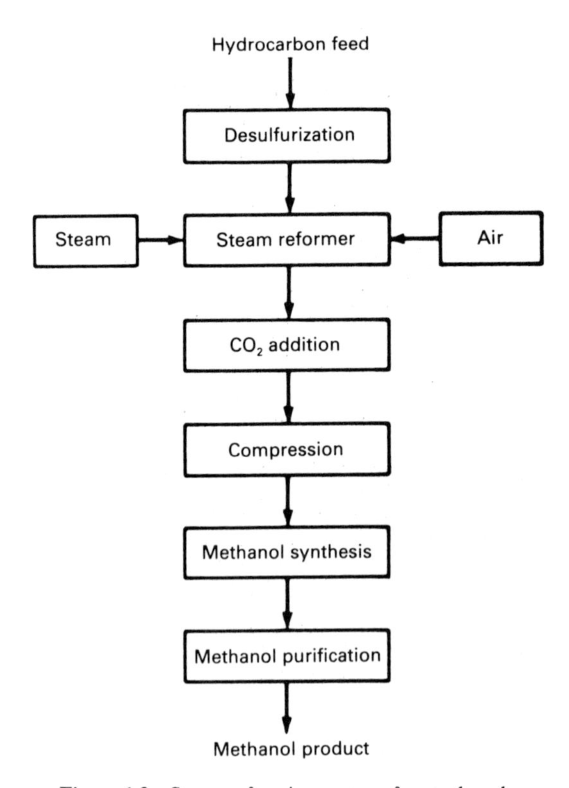

Figure 1.3 Steam reforming process for methanol

carbon fed to the partial oxidation reactors is unconverted and leaves the reactor as soot. The hot reactor effluent gas is quenched in waste heat boilers attached to each reactor, and high-pressure saturated steam is generated. Several commercial processes are available for the removal of the acid gases from the crude synthesis gas. The synthesis gas, after removal of the acid gases, has a hydrogen/carbon monoxide ratio of 1/1, which is converted to 2/1 by means of a conventional shift conversion and carbon removal step. The preceding steps are then followed by conventional methanol synthesis and purification, as described earlier. Figure 1-4 is a block flow diagram of a typical process.

Coal gasification processes have been used for many years to produce synthesis gas for steam generation and heating. Most of these processes were developed in Germany; typical examples of them are the Lurgi, Koppers-Totzek, and Winkler processes. The processes use air (or oxygen) and steam, and the synthesis gas has been used as raw material for ammonia, methanol, and synthetic liquid fuels in Europe but not in the United States. Substantial development has been under way in the past few years to improve the processes and permit production of an improved synthesis gas as feedstock for methanol and ammonia production. The gasifiers have, in general, been operated at higher pressures and temperatures. The higher pressure operation increases gasifier productivity and minimizes methane yield in the gasifier, which lowers production costs for the generation of the synthesis gas. Companies such as Texaco, Shell, Lurgi, and Hoechst have been active in the development. The Texaco gasifier has been successfully used by Tennessee Eastman in its project at Kingsport, Tennessee, for the production of methanol and acetic anhydride.

Research and development continues in the commercial synthesis of methanol. The last decade has seen a reduction in the pressure, a large reduction in energy requirements, and a tremendous increase in plant size. Recently attention is even being given to changes in the chemistry of the process. For example, researchers at Argonne National Laboratories in the United States have succeeded in producing methanol from carbon dioxide and water using a lead-based catalyst (2). Although the development of a viable process is in a very early stage, it could be important, particularly in the use of coal as feedstock.

Market

In the early eighties worldwide methanol demand was almost entirely for its use in chemical applications. About half of the methanol consumed is in the production of formaldehyde for use in various formaldehyde resins, such as urea, melamine, and thermoplastic acetyl resins. Virtually all formaldehyde is produced from methanol. Research work continues on the development of a commercial process for the manufacture of formaldehyde via direct oxidation of methane, but there is no indication that commercialization of such a process is close. Because of the major use of methanol for the production of formaldehyde, the successful commercialization of such a process would have a tremendous effect

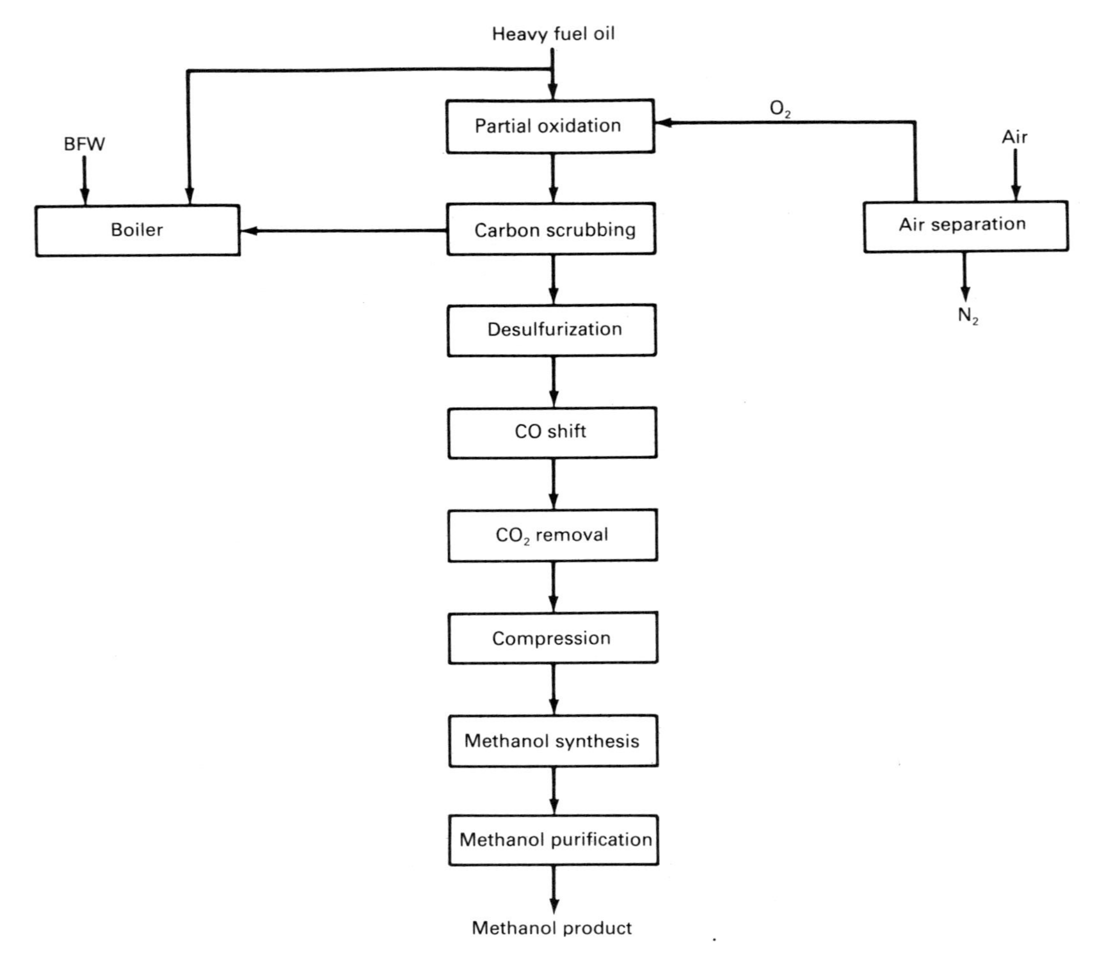

Figure 1.4 Partial oxidation process for methanol

on the world methanol business. Since formaldehyde consumption is primarily connected to the mature forest products industry, consumption is expected to grow at a fairly modest rate, perhaps 3 percent annually, primarily in those nations that are developing plywood, chipboard, and fiberboard industries. Another moderating factor is the ban on urea formaldehyde home insulation in some areas of the world because of possible health problems.

A much lower, but still substantial, use for methanol is in the production of acetic acid via a carbonylation reaction. The following decade is expected to see a phasing out of facilities using older technologies and their replacement with the carbonylation technology. This change will result in methanol requirements for acetic acid growing at a higher rate than the production of acetic acid per se. Thus although growth in acetic acid demand is anticipated to increase at perhaps a 2 percent rate during the next decade, demand for methanol for this use is expected to double in this same period of time.

Use of methanol in the production of dimethylterephthalate is projected to decrease over the next decade because of the gradual displacement of DMT with terephthalic acid as the preferred feedstock for the production of polyester fibers. Because of a production cost advantage in the use of the acid, the acid (which does not require methanol) has made continuous inroads into the market of the ester (which does require methanol), and this trend is expected to continue. The acid and ester are virtually inseparable with regard to their major application, which is the manufacture of polyester fibers and films.

Since most chemical markets for methanol are relatively mature, the growth in this area will parallel trends in economic activity. Worldwide methanol demand is projected to double in the next decade from the current figure of about 14 million metric tons. The largest single use for methanol by the end of the next decade is expected to be in liquid fuels. This application will include motor fuels, motor fuel additives, and fuel for power generation. Although pure methanol for motor fuel is being used in test vehicles, its large-scale use is developing slowly because of concerns over performance as well as stable crude oil prices, and it is not expected to be widely used in the short term.

Energy economics have changed dramatically in recent years. In the late seventies motorists in the United States and other oil dependent nations were contending with the second oil crisis. This crisis resulted in sharply curtailed supplies, substantially higher prices, and long waiting lines to purchase gasoline. This condition resulted in a panic situation that focused on the development of alternative fuels. The primary alternative, in the view of many, was methanol.

The situation in the past few years has been markedly different. With the world recession of the early eighties and the dramatic decrease in the world requirement for crude oil, a crude oversupply has resulted in a moderation in the cost of crude. There is, however, still substantial interest in pure methanol as fuel. Among the current sponsors in the United States of pure methanol motor fuels are Celanese, Ford Motor, the California Energy Commission, and Southland. The use of pure methanol will require engines designed specifically for its use and, in addition, the use of compatible materials of construction. Furthermore it would require suitable fuel distribution systems. These economic barriers are substantial and would probably require government subsidies. However, use of methanol in the manufacture of fuel additives such as methyl tertiary butyl ether is growing rapidly. Methyl tertiary butyl ether (MTBE) is an effective octane-improving additive for unleaded gasoline. One problem in the production of MTBE, however, is that it is limited by isobutylene availability. This limitation is due primarily to the worldwide shift to the use of ethane and LPG feedstocks in ethylene production, which produce less isobutylene than naphtha or gas oil does.

Probably the most important market for methanol in the next few years will be as a high-octane blending component in gasoline. However, this application is not without problems. Methanol is sensitive to moisture and is corrosive to conventional automobile components at concentrations over about 10 percent. Some automobile manufacturers have stated that use of methanol blends may jeopardize car owner warranties, and some oil companies have advertised a warning of car problems associated with methanol use. Methanol currently sells in the United States at about half the cost of unleaded gasoline on a volume basis but at about 80 percent on an equivalent Btu basis. However, it has an octane value of about 116 compared with 87 for most unleaded regular gasolines. The high octane is expected to be an especially important factor for refiners as most developed nations speed up the phaseout of lead in gasoline.

Another difficulty tending to delay the use of methanol in fuel is the opposition from a number of major oil companies that favor ethanol blending. Texaco, Chevron, and Amoco in the United States, for example, are major marketers of gasohol, and at least two of them have direct investments in ethanol-producing facilities.

Methanol is currently being blended with gasoline commercially in Europe and in the United States. The level of about 3 percent in Europe is primarily based on economic considerations, whereas in the United States the Environmental Protection Agency has set limits. In early 1985 the Environmental Protection Agency granted DuPont a waiver on the use of methanol in gasoline subject to certain conditions regarding concentration, cosolvents, corrosion inhibitors, and volatility (25). Properly blended, methanol in gasoline is said to reduce particulate emissions and levels of sulfur, nitrogen oxide, hydrocarbons, and carbon monoxide. In the United States Arco and Sun are the only major integrated refiners to use methanol. In Western Europe low-level methanol blends are being used in West Germany and Italy. Arco has introduced "Oxinol," an octane enhancer that combines equal parts of methanol and tertiary butyl alcohol, a cosolvent alcohol. Many of the problems associated with the use of methanol as a fuel are reportedly effectively limited with the use of a cosolvent alcohol. However, the unavailability of cosolvent alcohol is a serious obstruction to the further development of this market. Tertiary butyl alcohol is believed to be available only from Arco, and other C_3 and C_4 alcohols are quite expensive. As t-butyl alcohol use expands, additional producers will enter the market.

Many companies, including Snamprogetti, Anic, and Haldor Topsoe, have developed processes for converting synthesis gas into an alcohols mixture called MAS. This mixture contains methanol and required solubility improvers, which are usually higher alcohols. The product is said to be directly usable for gasoline blending because it substantially improves the behavior of gasoline-methanol blends with regard to phase separation and volatility problems. The processes are reportedly close to commercialization.

A synthetic gasoline process that produces methanol as an intermediate has been developed by Mobil Oil and is being commercialized in New Zealand and duplicated in other locations. The methanol produced in these facilities, however, is essentially destined for captive market outlets and will not affect the worldwide methanol business. The gasoline produced by this technology is high octane and free of sulfur and nitrogen. The methanol produced in New Zealand will use extensive supplies of natural gas as feedstock.

Additional nonfuel uses of methanol that will grow in the next decade include its role in the production of single-cell protein. ICI of the United Kingdom is actively pursuing this use. Production of ethylene and ethylene glycol and styrene from methanol via homolization and dehydration is not expected to be a major application during the next decade.

Natural gas and petroleum-derived hydrocarbons are expected to continue to be the major feedstock for many years. U.S. production will probably not be able to keep up with the demand but will remain the major market for methanol, taking perhaps a third of world output. Japan and Europe will probably account for about half, with the remainder distributed around the world. The distribution will result in gaps between demand and supply, and the difference will be made up by imports, probably from methanol plants in places such as Canada, Mexico, and the Middle East. The United States will probably join Western Europe and Japan as major methanol importers in a very short time. A result of these increased imports will undoubtedly be a narrowing of the price differential between relatively low U.S. prices and world prices.

Historically, methanol supply has been concentrated in the industrialized nations. In the early eighties the United States, Western Europe, and Japan accounted for almost two-thirds of the world supply. However, a large majority of the new methanol facilities being built worldwide are outside of the three major consuming regions. Large projects have been built, are under construction, or are in the planning stage in Canada, Mexico, Saudi Arabia, Trinidad, and Southeast Asia. In the early eighties worldwide methanol supply and demand were fairly well in balance, but since that time an excess of supply over demand has occurred and methanol from export-oriented countries has produced considerable downward pressure on prices.

Current interregional methanol trade is approximately 4 million metric tons and is expected to more than double within the next decade. The major exporting regions are expected to be Western Canada, the Middle East, and Eastern Europe. Additional supplies will come from Trinidad, New Zealand, and Southeast Asia. Western Europe will probably obtain its supplies from North Africa, the Middle East, and Eastern Europe; the U.S. supplies will come from Western Canada, Trinidad, and the Middle East. Japan will be the most dependent on imports and will obtain its material from joint ventures in Saudi Arabia as well as from New Zealand, Canada, and Malaysia.

As additional capacity comes on stream in the natural-gas-rich regions of the world, rationalization of capacity in the developed areas will be the rule. Substantial rationalization has already begun in Japan, Western Europe, and the United States. The influx of methanol is expected to force older, fuel-oil-based or naphtha-based methanol facilities to be shut down.

The shift of methanol production from the developed nations to the resources-rich countries with inexpensive natural gas has begun. In the United States no additional capacity has been proposed and several facilities have been shut down, including Celanese's facility at Clear Lake, Texas, and Tenneco's at Pasadena, Texas. Together these facilities account for over a million metric tons per year. DuPont has discontinued production of 675,000 metric tons per year of methanol at its Beaumont, Texas, facility as a result of worldwide overcapacity. These shutdowns are, of course, not irreversible and could be started up again if the supply and demand relationship and the prices improve significantly. Prices in the United States have dropped about a third since 1981, while production costs, particularly for natural gas, have continued to increase. At present 10 major producers, including the Deer Park, Texas, facility of DuPont, use heavy oil as raw material. All the other facilities use natural gas. DuPont withdrew from the merchant methanol market at the end of 1984. A summary of U.S. producers is shown in Table 1-4 (11).

Canada has three current producers: Alberta Gas Chemicals and Celanese Canada in Alberta, each of which has a capacity of about 650,000 metric tons per year, and Ocelot Industries in British Columbia, with a capacity of about 400,000 metric tons per year. The total capacity is about 1.7 million metric tons per year. Alberta Gas Chemicals has considered a facility at Scotford, Alberta, but as of this writing, work has not yet started and the future of the project is questionable. Biewag Energy Resources has proposed a very large facility in Alberta, but no target date has been set and the status of the plant is uncertain.

Total capacity for Latin America is about 750,000 metric tons per year, which includes 435,000 metric tons in Trinidad and another 160,000 metric tons in Mexico. Additional facilities are planned for Argentina, Chile, and Mexico, but their status is uncertain and they would not be in production until the late eighties, if at all.

Western European capacity is approximately 2.9 million metric tons per year, and no additional facilities are expected. On the contrary several facilities are expected to suspend operations in the next few years to compensate for

TABLE 1-4 Summary of U.S. Producers of Methanol

Producer	Location	Estimated Capacity mm lb/yr	m MT/yr
Air Products	Pensacola, Florida	397	180
Allemania Chemical	Plaquemine, Louisiana	858	390
Arco Chemical	Channelview, Texas	1323	600
Borden Chemical	Geismar, Louisiana	1386	629
Celanese	Bishop, Texas	992	450
	Clear Lake, Texas	1518	688
DuPont	Deer Park, Texas	1323	600
	Beaumont, Texas	1650	748
Georgia Pacific	Plaquemine, Louisiana	794	360
Texaco (Getty)	Delaware City, Delaware	662	300
Monsanto	Texas City, Texas	662	300
Tennessee Eastman[a]	Kingsport, Tennessee	430	195
	TOTAL	11,995	5440

[a]The Tennessee Eastman facility is used captively for the production of acetic anhydride. In addition, Tenneco has an 858 million pound per year facility at Pasadena, Texas, on standby.

imports from Eastern Europe, the Middle East, and Africa. Methanol from the USSR is supplied to Western Europe under a compensation contract and is a source of supply pressure. Table 1-5 indicates major Western European producers of methanol (12).

Eastern Europe is projected to be a major methanol producer in the next few years if facilities in the USSR and East Germany are completed. Two 800,000 metric ton per year facilities are planned in the USSR and a 700,000 metric ton

TABLE 1-5 Major Western European Producers of Methanol

Producer	Location	Estimated Capacity mm lb/yr	m MT/yr	Feedstock
Osterreichische Hiag.	Fischamend, Austria	165	75	
Soc. Methanolacq	Pardies, France	265	120	Natural gas
BASF	Ludwigshafen, W. Germany	529	240	Natural gas, naphtha
Chemische Werke Huls	Gelsenkirchen, W. Germany	441	200	Heavy oil
Union Rheinische	Wesseling, W. Germany	882	400	Heavy oil
Montedison	Castellanza, Italy	254	115	Natural gas
Methanor	Delfzijl, Netherlands	1632	740	Natural gas
Norsk Hydro	Porsgrunn, Norway	132	60	Heavy oil
CEPSA	Algeciras, Spain	441	200	Naphtha
ICI	Billingham, U.K.	1544	700	Natural gas
	TOTAL	6285	2850	

per year facility in East Germany. If these projects are completed, Eastern European capacity is expected to exceed that of the United States.

The Middle East, from virtually no capacity in the early eighties, will in the next few years become a major producer in the international methanol market. A 600,000 metric ton per year facility has already successfully started up in Saudia Arabia, and other Saudi facilities will be started up shortly. Bahrein is constructing a 330,000 metric ton per year facility, and the United Arab Emirates is considering a substantial capacity in the late eighties. As a result Middle East capacity is expected to match that of Canada in a few years.

For Asia and Oceania major additional methanol capacity is expected in Indonesia and Malaysia. A 330,000 metric ton per year facility is under construction in Indonesia, and a 660,000 metric ton per year facility is being built in Malaysia. A projection of worldwide capacity in the early nineties is shown in the following table:

Region		Estimated Capacity, thousand MT/yr
United States		5300
Canada		1600
Latin America (including Trinidad)		750
Western Europe		2900
Eastern Europe (including USSR)		5700
Middle East and Africa		1600
Asia and Oceania (including Japan)		2500
	TOTAL	20,250

Major current producers in Japan together with estimated capacities are shown in Table 1-6. Some of these facilities are expected to be shut down over the next few years. The first three producers use natural gas feedstock, the last two use butane feedstock.

Much of the new capacity that will be built is in countries with limited domestic demand, and the new producers will obviously have to export much of their production. The biggest impact is expected to be in Western Europe

TABLE 1-6 Major Japanese Producers of Methanol

Producer	Location		Estimated Capacity, thousand MT/yr
Mitsubishi Gas Chemicals	Niigata		200
Mitsui Toatsu Chemicals	Chiba		130
Kyowa Gas Chemical	Nakajo		130
Higashi-Nihon Methanol	Chiba		300
Nishi-Nihon Methanol	Himeji		300
		TOTAL	1060

and Japan as well as in the United States. Saudi Arabia will be a major source for Europe and Japan, whereas most of the imports to the United States will come from Canada and Trinidad. Clearly, the worldwide supplies of methanol will far exceed demand for the next decade unless major amounts of methanol end up in fuel. The key to this use is the price of crude oil, which will determine the future of methanol in fuel markets. If prices of crude remain low and stable, use of methanol in the fuel market will proceed slowly. If crude prices increase markedly, for whatever reason, the use of methanol in fuel will grow strongly and, in this case, additional capacity will undoubtedly be required before the end of the next decade.

Economics

Typical production costs for methanol from natural gas are shown in the following table. As with ammonia the major cost is from the natural gas, followed by overhead, the major part of which is depreciation.

Capacity: 730 mm lb/yr (331 m MT/yr)
Capital cost: BLCC, $82 mm; OSBL, $33 mm; WC, $24 mm

	¢/lb	$/MT	%
Raw materials[a]	6.2	137	62
Utilities	0.4	9	4
Operating costs	0.9	20	9
Overhead costs	2.5	55	25
Cost of production	10.0	221	100
Transfer price	14.7	324	

[a] Natural gas at $4/mm Btu

FORMALDEHYDE

Formaldehyde is one of the most versatile chemicals and is found in products from embalming fluids to toothpaste. It was a basic raw material for the early plastics industry. The original Bakelite is a thermosetting resin made from the reaction of formaldehyde and phenol. Development of other resins based on urea and melamine followed, and such products continue to be the main use for formaldehyde and a major factor in the steadily increasing demand for methanol, from which formaldehyde is commercially produced.

Technology

Via methanol. The principal route to formaldehyde is by the vapor-phase catalytic oxidation of methanol, although a small amount is still produced by the partial oxidation of lower hydrocarbons. The catalysts used are usually silver, copper, or iron-molybdenum oxide. Although many commercial processes are available, they do not differ greatly from each other except for the type of catalyst used. The process based on methanol involves passing the vapor together with air over a fixed bed of catalyst at approximately atmospheric pressure. The product gases are then absorbed in water. The formaldehyde is generally considered to be formed by two reactions, the dehydrogenation and the oxidation of methanol, as follows:

$$CH_3OH \longrightarrow HCHO + H_2$$

$$CH_3OH + \tfrac{1}{2}O_2 \longrightarrow HCHO + H_2O$$

The silver- or copper-catalyzed process generally operates with a rich methanol-air mixture and yields a primary product solution containing unreacted methanol. Conversions are generally about 65 percent, and yields are almost 90 percent. The iron-molybdenum-catalyzed process, on the other hand, generally operates with a lean methanol-air mixture and yields a substantially methanol-free product solution.

Formaldehyde users tend to be widely dispersed, and the usual procedure is to ship methanol, manufactured in large, central methanol facilities, to smaller formaldehyde plants. The costs of the formaldehyde plants are relatively small, and a large, central formaldehyde facility would require the shipping of the usual aqueous formaldehyde solution (37 percent).

The feedstock of choice for formaldehyde will continue to be methanol, for the short term at least. Methanol is currently manufactured from several feedstocks in the various producing countries. In the United States it is made primarily from synthesis gas produced from natural gas or heavy-hydrocarbon mixtures. A block flow diagram of a typical process is shown in Figure 1-5.

Via oxidation of methane. Production of formaldehyde from natural gas via methanol involves three steps: (1) production of synthesis gas from natural gas, (2) production of methanol from synthesis gas, and (3) production of formaldehyde from methanol. Not surprisingly a substantial effort has been expended to oxidize methane directly to formaldehyde. Thus far, however, the

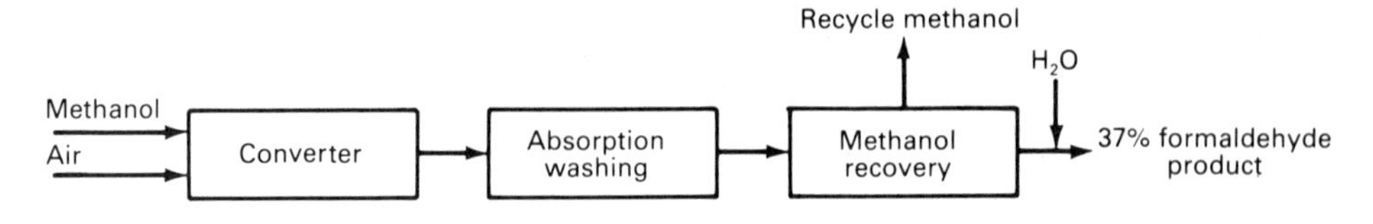

Figure 1.5 Production of formaldehyde from methanol

difficulties of attaining economical yields and productivity levels have been overwhelming. Methane is more resistant to oxidation than formaldehyde is, and the formaldehyde is oxidized relatively rapidly to carbon dioxide and water at temperatures necessary for reasonable reaction rates. Most of the literature shows low conversions and short residence times to produce reasonable selectivity to formaldehyde. As an example a relatively recent patent issued to Chemische Werke Huels of West Germany describes a specially designed reactor and a control of gas velocities and report mixtures of formaldehyde and methanol of up to 80 percent at 3 percent methane conversion (2). Higher conversions reduce yields. The patent describes the use of air or oxygen, but the low conversion and the need to recycle large quantities of methane would seem to eliminate the use of air. The need for oxygen and the control problems in the high-pressure (40 atmospheres) reactor indicates that commercialization is still a long way in the future.

Market

Most of the formaldehyde in the United States is sold as a 37 percent aqueous solution. Demand in 1983 increased to about 5.7 billion pounds, after dropping to less than 5 billion pounds in 1982, and remained relatively steady in 1984. High demand occurred in 1979, when a level of 6.4 billion pounds was reached. The increase in production was due primarily to the recovery of the housing industry, which involves formaldehyde in several major resins used in adhesives and plastics. Urea-formaldehyde resins account for about 27 percent of the market in the United States, and phenol-formaldehyde resins take about another 20 percent. Other major uses are for 1,4-butanediol and polyacetal resins, which together account for about 15 percent of the market. The remainder of the market, is distributed mainly for the production of pentaerythritol, hexamine, melamine, and MDI. There is still substantial overcapacity, and therefore new installations are not expected to be built for several years. Major U.S. producers of formaldehyde are shown in Table 1-7 (4,5).

Many formaldehyde producers exist throughout Western Europe, but a large fraction of the formaldehyde is produced in West Germany by companies including BASF, Bayer, and Degussa. Table 1-8 shows most of the major producers, divided by country (12). With regard to feedstock, no problem is anticipated. Although demand for fuel alcohol and other uses will increase, new methanol capacity worldwide should satisfy the demand quite easily.

Japan has at least a dozen manufacturers of formaldehyde in various forms and solutions, including several major chemical companies. These companies include Mitsubishi Chemical Industries, Mitsubishi Gas Chemical, Kyowa Gas Chemical Industry, Mitsui Toatsu Chemicals, Sumitomo Chemical, and Dainippon Ink and Chemicals.

TABLE 1-7 Major U.S. Producers of Formaldehyde

Producer	Locations	Estimated Capacity As 37% Solution	
		mm lb/yr	m MT/yr
Borden (11 facilities)	Alabama, N. Carolina, Oregon, Kentucky, Wisconsin, Montana, Washington, Texas, Louisiana, California	1600	726
Celanese (3 facilities)	Texas, New Jersey, S. Carolina	1900	862
DuPont (5 facilities)	Texas, Ohio, New Jersey, W. Virginia, N. Carolina	1500	680
GAF (2 facilities)	Texas, Kentucky	200	91
Georgia Pacific (8 facilities)	Oregon, Ohio, Arkansas, S. Carolina, N. Carolina, Mississippi, Georgia, Texas	1100	499
Hercules	Missouri	170	77
International Min.	Pennsylvania	135	61
Monsanto (4 facilities)	Ohio, Texas, Oregon, Massachusetts	700	317
Nuodex	New Jersey	185	84
Perkins Industries	Mississippi	60	27
Reichhold (6 facilities)	S. Carolina, Texas, Kansas, Washington, Alabama, Oregon	600	272
Wright Chemical	N. Carolina	80	36
	TOTAL	8230	3732

TABLE 1-8 Major Western European Producers of Formaldehyde

Producer	Location	Estimated Capacity, As 37% Solution	
		mm lb/yr	m MT/yr
Krems Chemie	Krems-An-Der-Donau, Austria	154	70
Oster. Hiag-Werke	Vienna, Austria	298	135
Other Austria	Austria	90	41
Other Belgium	Belgium	77	35
Nordalim	Arhus, Denmark	154	70
Oy Noresin	Kitee, Finland	132	60
Priha Oy	Hamina, Finland	238	108
Cegecol	Ambares-et-LaGrave, France	99	45
Organichim	Mazingarbe, France	146	66
	Premery, France	84	38
	Villers-Saint-Paul, France	298	135

TABLE 1-8 Major Western European Producers of Formaldehyde (*continued*)

Producer	Location	Estimated Capacity, As 37% Solution	
		mm lb/yr	m MT/yr
Hoechst	Trosly-Breuil, France	71	32
SNPE	Toulouse, France	108	49
STS	Toulouse, France	146	66
Other France	France	24	11
Bakelite	Duisburg, W. Germany	110	50
BASF	Ludwigshafen, W. Germany	2403	1090
Bayer	Krefeld, W. Germany	269	122
	Leverkusen, W. Germany	298	135
Chemische Werke	Perl, W. Germany	280	127
Degussa	Arnsberg, W. Germany	714	324
GAF-Huls Chemie	Marl, W. Germany	154	70
Ticona Polymerwerke	Kelsterbach, W. Germany	392	178
Other W. Germany	W. Germany	260	118
Victor	Sindos, Greece	110	50
Other Greece	Greece	49	22
Chimica Pomponesco	Pomponesco, Italy	110	50
Fabrica Adesivi Res.	Filago, Italy	220	100
Montedison Resine	Castellanza, Italy	838	380
Sadepan Chimica	Viadana, Italy	220	100
SIR	Solbiate, Italy	287	130
SPREA	Castelseprio, Italy	110	50
Other Italy	Italy	110	50
Methanol Chemie	Delfzijl, Netherlands	357	162
	Rozenburg, Netherlands	329	149
Dyno Industrier	Lillestrom, Norway	176	80
	Saetre-I-Huram, Norway	209	95
Bresfor	Gafanha-Da-Nazare, Portugal	110	50
SARL	Maia, Portugal	88	40
Borden	Axpe-Erandio, Spain	143	65
Derivados Forestales	San-Celoni, Spain	198	90
FORESA	Caldas-De-Reyes, Spain	220	100
Formol y Derivados	Almusafes, Spain	287	130
Other Spain	Spain	57	26
AB Casco	Kristinehamn, Sweden	220	100
	Sundsvall, Sweden	110	50
Perstorp Kemi	Perstorp, Sweden	375	170
Other Switzerland	Switzerland	99	45
BIP Chemicals	Oldbury, U.K.	132	60
Borden	Peterlee, U.K.	88	40
Ciba-Geigy	Duxford, U.K.	265	120
Synthite	West Bromwich, U.K.	249	113
Other U.K.	U.K.	159	72
	TOTAL	12,924	5872

Toxicity

In the early eighties a major problem for formaldehyde producers developed regarding the toxicity of formaldehyde, particularly of urea formaldehyde foam used in home insulation (7). Toxicity research on formaldehyde has expanded rapidly over the past few years. The U.S. Formaldehyde Institute, an association of producers, is concerned that the ban on foam could be a first step in a wider series of bans on products that give off formaldehyde vapor. Of particular concern are products used in textiles and particle board and plywood used in home construction.

Economics

Approximate production costs for a typical formaldehyde facility are shown in the following table:

Capacity: 110 mm lb/yr (50 m MT/yr)
Capital cost: BLCC, $8.5 mm; OSBL, $3.4 mm; WC, $2.5 mm

	¢/lb	$/MT	%
Raw materials[a]	3.6	80.3	59
Utilities	0.5	10.3	7
Operating costs	0.5	10.3	7
Overhead costs	1.7	37.3	27
Cost of production	6.3	138.2	100
Transfer price	9.5	209.8	

[a] Methanol at 54¢/gal

ACETIC ACID

The production of acetic acid has three major commercial routes: (1) the oxidation of acetaldehyde; (2) the oxidation of paraffin hydrocarbons, mainly normal butane; and (3) the carbonylation of methanol.

Technology

Methanol carbonylation is generally considered to be the most efficient route to acetic acid. Oxidation of *n*-butane capacity is decreasing, since many units using this process have been placed on standby (2). Most of these facilities are fully depreciated, and their coproduct values will keep them economically viable for some years to come. Acetaldehyde oxidation has been losing ground rapidly because of increasing costs for ethylene, the feedstock for production of the acetaldehyde. The methanol carbonylation process is anticipated to replace this process if additional capacity is required in the future. Approximately half of the capacity in the United States is currently using the methanol carbonylation process, and this figure is expected to increase in the future.

The methanol carbonylation process involves the reaction of liquid methanol and carbon monoxide under mild conditions of temperature and pressure. A rhodium catalyst system is used. Only small amounts of by-products are formed, with the yields of acetic acid on methanol exceeding 99 percent and the yields on carbon monoxide exceeding 90 percent. The reaction is as follows:

$$CH_3OH + CO \longrightarrow CH_3COOH$$

The secondary reaction is:

$$CO + H_2O \longrightarrow CO_2 + H_2$$

The acetic acid is withdrawn from the reaction system and purified by conventional distillation. A block flow diagram for the process is shown in Figure 1-6.

Methanol and carbon dioxide are fed continuously to a liquid-phase reactor containing acetic acid, water, hydrogen iodide, methyl iodide, and the rhodium catalyst. The methanol reacts rapidly with hydrogen iodide to form methyl iodide, which reacts with the rhodium complex and carbon monoxide to yield acetic acid.

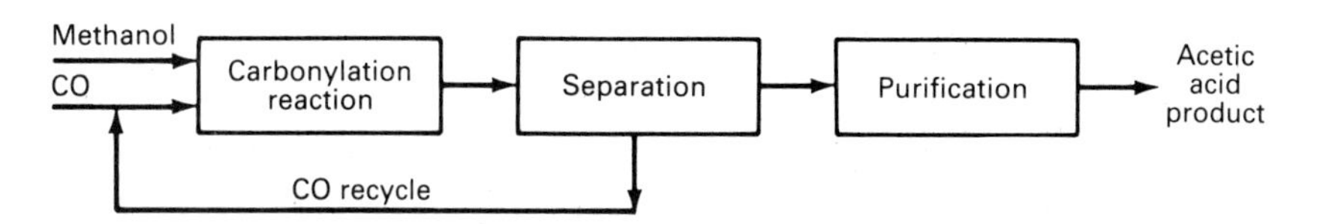

Figure 1.6 Production of acetic acid by methanol carbonylation

Market

Acetic acid is a well-established basic organic chemical, and relatively modest growth of perhaps 3 percent a year is expected through the eighties. In the past few years, almost a billion pounds per year of acetaldehyde oxidation capacity have been shut down, and other inefficient facilities are expected to close in the eighties. Production in the United States in 1983 was about 2.7 billion pounds, and the projection by 1987 is generally expected to be about 3.2 billion pounds (3).

Growth in acetic acid demand will largely depend on the markets for vinyl acetate and terephthalic acid. Vinyl acetate is by far the largest market for acetic acid and accounts for almost half the demand. Vinyl acetate monomer is used for paints, coatings, and adhesives and is therefore tied to the construction industry, which in turn depends on the general state of the economy. In the early eighties demand was poor because of the worldwide recession, but a pickup in demand began in 1984. The start-up in 1984 of Eastman's acetic-anhydride-from-coal facility in Tennessee will reduce that company's acetic acid requirements by about 100 million pounds. Eastman has a 630 million pound per year ethylene-based acetic acid facility at the same location. Once the new unit is up to production capacity, the existing acetic acid facility is expected to be used for captive purposes and only for augmentation of the coal facility.

Acetic acid is also used as a solvent in the manufacture of terephthalic acid, and this use is expected to grow. The terephthalic acid is used in turn for the production of polyethylene terephthalate, which is used in the manufacture of polyester fiber and in beverage bottles. This latter sector is expected to grow substantially during the balance of the eighties.

Other uses for acetic acid are relatively minor. These uses include production of acetic anhydride for materials other than cellulose acetate, acetic esters for paints and coatings, textile applications, and monochloracetic acid for herbicides and carboxymethyl cellulose. The approximate proportion of acetic acid used for the different end products are summarized in Table 1-9.

Relatively small amounts of acetic acid are, and will probably continue to be, exported as acetaldehyde oxidation facilities outside the United States are phased out because of poor economics attributed to rising ethylene costs. On the other hand exports will be threatened as new capacity for both acetic acid and vinyl acetate comes on stream overseas. U.S. exports go primarily to Western Europe, Japan, and other Pacific Rim countries. Major U.S. producers with their approximate capacities are shown in Table 1-10 (2,7).

In addition to the producers listed in Table 1-10, Monsanto and Air Products at Springfield, Massachusetts, produce small amounts of acetic acid from polyvinyl alcohol production; Union Carbide at Taft, Louisiana, produces about 40 million pounds per year as a by-product from peracetic acid production; and DuPont at LaPorte, Texas, has the capability to produce over 100 million pounds per year from polyvinyl alcohol production. Air Products will increase its capacity at Calvert City, Kentucky, by 50 million pounds per year by the end of 1985. Publicker has an 80 million pound per year facility at Philadelphia, which has been shut down for several years.

Major producers in Western Europe are shown in Table 1-11 (5). In Japan

TABLE 1-9 Approximate Percentage of Acetic Acid Used for End Products

End Use	Approximate %
Vinyl acetate	45
TPA	12
Cellulose acetate	25
Other	18

TABLE 1-10 Major U.S. Producers of Acetic Acid

Producer	Location	Estimated Capacity	
		mm lb/yr	m MT/yr
Air Products	Calvert City, Kentucky	75	34
Borden[a]	Geismar, Louisiana	150	68
Celanese	Bay City, Texas[b]	300	136
	Clear Lake, Texas[c]	800	363
	Pampa, Texas[d]	550	249
Eastman[e]	Kingsport, Tennessee	630	286
Monsanto[f]	Texas City, Texas	400	181
USI[g]	Deer Park, Texas	600	272
Union Carbide[h]	Brownsville, Texas	700	317
	TOTAL	4205	1906

[a] From methanol on standby 1984

[b] Wacker acetaldehyde process; on standby 1984

[c] From methanol

[d] From *n*-butane

[e] Wacker acetaldehyde process; expected to be phased out when chemicals from coal project are on full production

[f] From methanol

[g] From methanol

[h] From *n*-butane; on standby 1984

nine acetic acid producers are currently in operation, including Chisso, Daicel Chemical Industries, Kyowa Hakko Kogyo, Nippon Synthetic Chemical Industry, Showa Acetyl Chemicals, Tokuyama Petrochemical, and Toyo Soda.

Economics

Typical production costs for the manufacture of acetic acid from methanol and carbon monoxide are as follows:

Capacity: 330 mm lb/yr (150 m MT/yr)
Capital cost: BLCC, \$46 mm; OSBL, \$20 mm; WC, \$18 mm

	¢/lb	\$/MT	%
Raw materials[a]	10.0	221	63
Utilities	2.0	44	13
Operating costs	0.8	18	5
Overhead	3.0	66	19
Cost of production	15.8	349	100
Transfer price	21.8	481	

[a] Methanol at 54¢/gal; carbon monoxide at 9¢/lb

TABLE 1-11 Major Western European Producers of Acetic Acid

Producer	Location	Estimated Capacity		Raw Material
		mm lb/yr	m MT/yr	
Rhone-Poulenc	LePont-de-Claix, France	55	25	Methanol
	Pardies, France	496	225	Methanol
BASF	Ludwigshafen, W. Germany	99	45	Methanol
Chemische Werke Huls	Marl, W. Germany	88	40	Methanol
Hoechst	Frankfurt, W. Germany	320	145	*n*-butene
	Hurth-Knapsack, W. Germany	176	80	Acetaldehyde
Wacker Chemie	Burghauser, W. Germany	176	80	Acetaldehyde
Montedison	Porto-Marghera, Italy	132	60	Acetaldehyde
Vinavil	Villadossola, Italy	132	60	Acetaldehyde
Akzo Zout Chemie	Rozenburg, Netherlands	221	100	*n*-butane
Ind. Quim. Asoc.	Tarragona, Spain	221	100	Acetaldehyde
Lonza	Visp, Switzerland	99	45	
BP Chemicals	Hull, U.K.	772	350	Naphtha, methanol
	TOTAL	2987	1355	

ACETIC ANHYDRIDE

Acetic anhydride is commercially produced through many routes. In the United States the processes in greatest use are the ketene-based technology, which uses the dehydration of acetic acid or the decomposition of acetone as the starting point, and acetaldehyde oxidation, which produces both acetic acid and acetic anhydride. The largest U.S. facility belongs to Tennessee Eastman at its Tennessee chemicals-from-coal complex, which employs a new process to produce acetic anhydride.

Technology

Ketene process from acetic acid. This technology involves pyrolysis of glacial acetic acid in the presence of a small amount of triethyl phosphate. The reaction takes place at 700 to 800°C and reduced pressure as follows:

$$CH_3COOH \longrightarrow CH_2{=}C{=}O + H_2O$$

$$CH_3COOH + CH_2{=}C{=}O \longrightarrow (CH_3CO)_2O$$

Overall yields are in the range of 85 to 90 percent. Figure 1-7 shows a block flowsheet for this process.

Ketene process from acetone. With acetone pyrolysis occurs at a lower temperature than with acetic acid, about 650°C, with yields in the range of 85 to 90 percent in accordance with the following reactions:

$$CH_3COCH_3 \longrightarrow CH_2{=}C{=}O + CH_4$$

$$CH_2{=}C{=}O + CH_3COOH \longrightarrow (CH_3CO)_2O$$

Acetaldehyde oxidation. This technology proceeds via the intermediate production of peracetic acid using a cobalt acetate catalyst. Temperatures are about 50°C and pressures about 50 psi. A diluent such as ethyl acetate or acetic acid is commonly used. Air or oxygen can be used, but the trend is toward

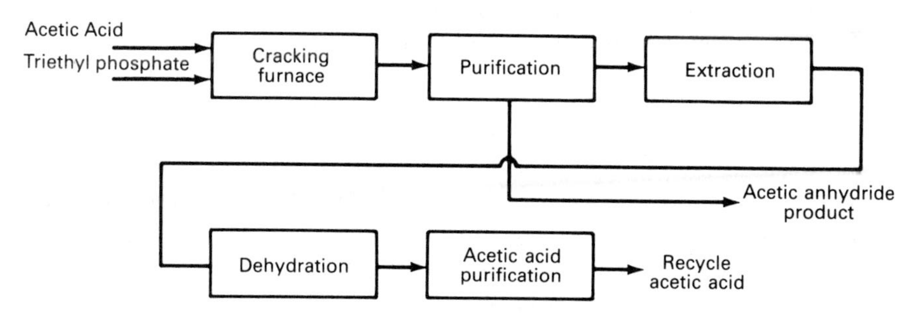

Figure 1.7 Production of acetic anhydride—ketene process from acetic acid

oxygen, often injected at points along the reactor tube. This process results in the production of equal weights of acetic acid and acetic anhydride. Conversion of acetaldehyde is about 95 percent.

$$CH_3CHO + O_2 \longrightarrow CH_3COOOH$$

$$CH_3COOOH + CH_3CHO \longrightarrow (CH_3CO)_2O + H_2O$$

$$CH_3COOOH + CH_3CHO \longrightarrow 2CH_3COOH$$

Acetic anhydride from coal. This process is a combination of technologies developed by Eastman Kodak and Halcon International and is used at an Eastman complex at Kingsport, Tennessee. The facility involves the following series of steps:

1. Production of synthesis gas from coal.
2. Production of methanol from synthesis gas.
3. Reaction of methanol and acetic acid to produce methyl acetate.
4. Carbonylation of methyl acetate to produce acetic anhydride.

The key reaction, carbonylation of methyl acetate, occurs at about 175°C using a rhodium chloride and chromium hexacarbonyl catalyst system in acetic acid with methyl iodide as a promoter. Conversion is said to be over 75 percent with 95 percent selectivity according to the following reaction:

$$CH_3COOCH_3 + CO \longrightarrow (CH_3CO)_2O$$

The reaction products appear to be a function of the hydrogen/carbon monoxide ratio. As the hydrogen content of the system is increased, there is a corresponding increase in ethylidene diacetate, the vinyl acetate precursor. Conversely, acetic anhydride yield is increased as hydrogen content is decreased.

Market

Acetic anhydride is a very large consumer of acetic acid and is primarily used in the production of cellulose acetate. Most of the remainder is used in the production of aspirin and other pharmaceuticals and pesticides. Because this material is closely tied to cellulose acetate and the growth rate for this material is low, the demand for acetic anhydride is generally expected to be essentially constant in the near future at a level of about 1.7 billion pounds in the United States. Since the start-up of Eastman's facilities, ample capacity should exist for many years. Major U.S. producers of acetic anhydride are shown in Table 1-12 (2,3).

In addition to the producers listed in Table 1-12, Carbide has a 200 million pound per year facility at Brownsville, Texas, which is on standby, and Avtex

TABLE 1-12 Major U.S. Producers of Acetic Anhydride

Producer	Location	Estimated Capacity	
		mm lb/yr	m MT/yr
Celanese	Pampa, Texas	300	136
	Narrows, W. Virginia	250	113
	Rock Hill, S. Carolina	250	113
Eastman	Kingsport, Tennessee	1500	680
	TOTAL	2300	1042

has a 60 million pound per year facility at Meadville, Pennsylvania. The Celanese facility at Pampa, Texas, is the only one now operating that produces virgin anhydride exclusively. Other than the Eastman facility, which produces the material from coal, the other plants recycle material that is produced in the manufacture of cellulose acetate.

Major producers of acetic anhydride in Western Europe are shown in Table 1-13 (6). The three major acetic anhydride producers in Japan are shown in Table 1-14.

TABLE 1-13 Major Western European Producers of Acetic Anhydride

Producer	Location	Estimated Capacity	
		mm lb/yr	m MT/yr
Rhone-Poulenc	Roussillon, France	150	68
Hoechst	Hurth-Knapsack, W. Germany	175	79
Lonza-Werke	Waldshut, W. Germany	18	8
Wacker-Chemie	Burghausen, W. Germany	33	15
Vinavil	Villadossola, Italy	125	57
Lonza	Visp, Switzerland	15	7
Courtaulds Acetate	Coventry, U.K.	500	227
	TOTAL	1016	461

TABLE 1-14 Major Japanese Producers of Acetic Anhydride

Producer	Location	Estimated Capacity	
		mm lb/yr	m MT/yr
Chisso	Minamata	15	7
Daicel Chem. Ind.	Amihoshi	265	120
	Arai	33	15
Nippon Syn. Chem. Ind.	Ohgaki	9	4
	TOTAL	322	146

Economics

Typical production costs for acetic anhydride from acetic acid via the ketene process are as follows:

Capacity: 300 mm lb/yr (136 m MT/yr)
Capital cost: BLCC, $119 mm; OSBL, $60 mm; WC, $49 mm

	¢/lb	$/MT	%
Raw materials[a]	33.2	732	68
Utilities	6.0	132	12
Operating costs	1.8	40	4
Overhead costs	8.1	179	16
Cost of production	49.1	1083	100
Transfer price	17.9	1478	

[a] Acetic acid at 25¢/lb

METHYL TERTIARY BUTYL ETHER

Methyl tertiary butyl ether (MTBE) is a widely accepted octane booster for unleaded gasolines. Although demand is expected to grow, many new facilities have been built and capacity is currently running ahead of demand.

Technology

MTBE is generally produced by the reaction of methanol with isobutylene as follows:

$$CH_3-\underset{\displaystyle CH_3}{\overset{\displaystyle CH_3}{\overset{|}{\underset{|}{C}}}}=CH_2 + CH_3OH \longrightarrow CH_3-\underset{\displaystyle CH_3}{\overset{\displaystyle CH_3}{\overset{|}{\underset{|}{C}}}}-OCH_3$$

The reaction takes place in the liquid phase using an acidic ion exchange catalyst. Temperatures are below 100°C, and pressures are relatively low. If a mixed C_4 stream is used, the isobutylene reacts and the other components are virtually unchanged. Very high yields are obtainable if excess methanol is used. The methanol and MTBE can be separated into an overhead azeotrope, which is recycled, and pure MTBE as bottoms. A mixture of MTBE and methanol is a possible marketing product. Small quantities of by-products such as diisobutene and tertiary butyl alcohol are generated.

The present source of isobutylene is almost entirely from refinery cracking units, either steam or catalytic. If the market grows sharply, substantial additional quantities of isobutylene will be required. These quantities might be obtained by isomerization of normal butenes or by dehydrogenation of isobutanes. Commercial processes for the isomerization of normal butane are available. An example is the BP process, in which the normal butane is isomerized to isobutane over a platinum-containing catalyst in the presence of hydrogen. The Catofin process, licensed by Houdry, could be used for the dehydrogenation of the isobutane. The Catofin catalyst is in the form of cylindrical pellets, consisting of activated alumina impregnated with chromic oxide. A block flow diagram is shown in Figure 1-8.

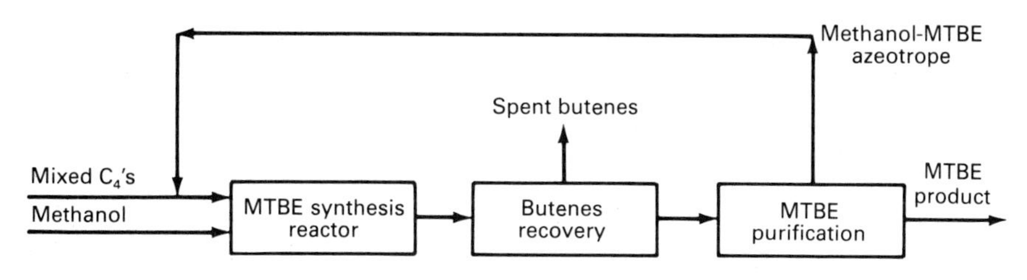

Figure 1.8 Production of MTBE from mixed C_4's

In the initial step fresh liquid makeup of C_4's, fresh methanol feed, and recycled methanol-MTBE azeotrope are preheated against the reactor effluent. The reaction is exothermic, and very high isobutene conversion per pass and selectivity to MTBE are realized. The reactor generally is a multitube shell exchanger. The tubes are packed with the catalyst pellets, and the heat of reaction is removed by circulating cooling water in the shell. A packed bed reactor can also be used, as can a combination of the two reactors.

Market

The use of lead compounds as an octane improver is being phased out in the United States, Canada, and Japan and also in many countries in Europe. Oxygenated compounds, including MTBE, are currently being used in many sections of the world. Primary interest in MTBE is in Europe and the United States. Although MTBE has been available for gasoline blending in Europe for many years, approval in the United States dates to February 1979, when the U.S. Environmental Protection Agency approved concentrations in gasoline of up to 7 volume percent. In late 1980 the rule was modified to allow a maximum of 2 weight percent of oxygen in unleaded gasoline. The compound was first manufactured by Chemische Werke Huels, in West Germany and ANIC in Italy.

Production of MTBE can be attractive for those refiners that have either a special economic incentive, such as low-cost methanol, or a petrochemical incentive to separate isobutylene from mixed butylenes. MTBE can also be attractive as a component to replace toluene in gasoline for use as a petrochemical feedstock.

TABLE 1-15 Major U.S. Producers of MTBE

Producer	Location	Estimated Capacity		Remarks
		mm lb/yr	m MT/yr	
Arco Chemical	Channelview, Texas	440	200	Own technology
GHR Energy	Good Hope, Louisiana	260	118	Huels technology
Tenneco	Houston, Texas	610	277	Snamprogetti technology
Phillips	Sweeney, Texas	260	118	Own technology
Charter	Houston, Texas	130	59	Chemi Research technology
Texaco	Port Neches, Texas	600	272	Deutsche Texaco technology
Exxon	Baytown, Texas	230	104	Huels technology
Champlin	Corpus Christi, Texas	150	68	Huels technology
	TOTAL	2680	1216	

TABLE 1-16 Major Western European Producers of MTBE

Producer	Location	Estimated Capacity		Remarks
		mm lb/yr	m MT/yr	
OMV	Schwechat, Austria	110	50	
NV SIBP	Antwerp, Belgium	221	100	
Neste Oy	Kulloo, Finland	176	80	Snamprogetti technology
Chem. Werke Huels	Marl, W. Germany	331	150	
Deutsche Texaco	Heide, W. Germany	26	12	
Enichem Polimeri	Ravenna, Italy	221	100	
DSM	Geleen, Netherlands	165	75	Snamprogetti technology
Shell	Rotterdam, Netherlands	243	110	Own technology
	TOTAL	1493	677	

Major U.S. producers of MTBE are shown in Table 1-15 (8). Major producers in Western Europe are shown in Table 1-16 (9). Facilities have also been built in Yugoslavia, Czechoslovakia, and East Germany as well as in Japan and South America.

Note that refiners have several competing options for boosting octane numbers. They can use more aromatics; increase the amount of butylene or propylene alkylate; or add tertiary butyl alcohol, methanol, ethanol or other alcohols. The choice of options is a matter of overall economics and must be weighed against the availability of isobutylene.

Economics

Typical production costs for the manufacture of MTBE from steam cracker C_4's are shown in the following table. Note that the prices assigned to the mixed C_4's and the by-product butenes have the major effect on the cost of production.

Capacity: 300 mm lb/yr (136 m MT/yr)
Capital costs: BLCC, $4.3 mm; OSBL, $5.6 mm; WC, $14.0 mm

	¢/lb	$/MT	%
Raw materials[a]	30.6	675	180
Utilities	0.6	13	4
Operating costs	0.2	4	1
Overhead costs	1.0	23	6
By-product credit[a]	(15.4)	(340)	−91
Cost of production	17.0	375	100
Transfer price	20.3	448	

[a] Mixed C_4's and spent butenes at 18¢/lb; methanol at 54¢/gal

REFERENCES

Ammonia

1. *1984 Directory of Chemical Producers—United States,* SRI International.
2. *1984 Directory of Chemical Producers—Western Europe,* SRI International.
3. "Key Chemicals—Ammonia," *Chemical and Engineering News,* February 4, 1985.
4. "The Ammonia Crunch," *Chemical Week,* February 23, 1983, pp. 34–38.
5. Omori, T. "Feedstock Switchover to Coal in Ammonia Production," *Chemical Economy and Engineering Review,* September 1982, Vol. 14, No. 9 (No. 161), pp. 7–10.
6. Esaki, M. "Trends in Ammonia Plants in Southeast Asia," *Chemical Economy and Engineering Review,* September 1982, Vol. 24, No. 9 (No. 161), pp. 11–21.
7. "West European Ammonia Deficiency Predicted to Increase by 1985," *European Chemical News,* October 4, 1982, pp. 16.

8. Brown, A. S. "Is the Ammonia Cycle Obsolete," *Chemical Marketing Reporter,* November 17, 1980, pp. 8–16.
9. *Kirk-Othmer Encyclopedia of Chemical Technology,* Third Edition, Vol. 2. New York: Wiley, 1984 pp. 470–516.
10. "Petrochemical Handbook '83," *Hydrocarbon Processing,* November 1983.
11. *JCW Chemicals Guide 82/83.* Japan: The Chemical Daily Co. Ltd. March 1982.
12. *Chemical Industry Yearbook,* Second Edition. Surrey, England: Industrial Press, 1984.
13. "Europe Can Compete in Ammonia Market," *European Chemical News, Fertilizers and Agrochemicals Supplement,* February 18, 1985, pp. 4–8.

Methanol

1. List, H. L. "Special Issue—Methanol, The Next Decade," *International Petrochemical Developments,* Vol. 5, No. 22, November 15, 1984.
2. "The New Route from CO to Methanol—via Water," *Chemical Week,* October 10, 1984, pp. 50–55.
3. "Large-Volume Fuel Market Still Eludes Methanol," *Chemical and Engineering News,* July 16, 1984, pp. 9–16.
4. "Methanol Supply to Outstrip Demand Until at Least 1990," *European Chemical News,* December 19/26, 1983, pp. 10–12.
5. Uhl, W. C. "More Hitches in Methanol's Growth Plan," *Chemical Business,* June 1984, pp. 27–36.
6. Kobayashi, K. "International Trends in Methanol," *Chemical Economy and Engineering Review,* June 1984, Vol. 16, No. 6 (No. 179), pp. 32–38.
7. "Methanol Touted As Best Alternate Fuel for Gasoline," *Chemical and Engineering News,* June 11, 1984, pp. 14–16.
8. "Why Europe Worries About Methanol," *Chemical Week,* December 14, 1983, pp. 22–24.
9. "Key Motor Fuel Role Seen for Methanol," *Oil and Gas Journal,* May 16, 1983, pp. 37–39.
10. "Why Celanese Is Pushing Methanol As an Auto Fuel," *Chemical Week,* September 21, 1983, pp. 41–44.
11. *1984 Directory of Chemical Producers—United States,* SRI International.
12. *1984 Directory of Chemical Producers—Western Europe,* SRI International.
13. List, H. L. "Methanol—The Next Decade," *International Petrochemical Developments,* Vol. 2, No. 4, February 15, 1981.
14. "Petrochemical Handbook '83," *Hydrocarbon Processing,* November 1983.
15. *Kirk-Othmer Encyclopedia of Chemical Technology,* Third Edition, Vol. 15. New York: Wiley, 1984, pp. 398–414.
16. "Chemical Profile," *Chemical Marketing Reporter,* September 19, 1983.
17. "Global Methanol Overcapacity Will Get Worse," *Chemical and Engineering News,* June 20, 1983, pp. 20–21.
18. "Emerging Energy and Chemical Applications of Methanol: Opportunities for Developing Countries," *World Bank Report,* 1982.

19. Uematsu, S. "Methanol Technology and Demand Trends and Their Future Prospects," *Chemical Economy and Engineering Review,* April 1983, Vol. 15, No. 4 (No. 167), pp. 5–13.
20. "Future Methanol Market Hangs on Potential for Fuel," *European Chemical News,* Petrochemicals '82 Supplement, December 20, 1982.
21. Unzelman, G. H. "Problems Hinder Full Use of Oxygenates in Fuel," *Oil and Gas Journal,* July 2, 1984, pp. 59–65.
22. *JCW Chemicals Guide 82/83.* Japan: The Chemical Daily Co. Ltd. March 1982.
23. "Methanol Producers See Nowhere Else to Go but Up, Though Imports Are a Problem," *Chemical Marketing Reporter,* February 4, 1985.
24. Cohen, L. H. and H. L. Muller, "Methanol Cannot Economically Dislodge Gasoline," *Oil and Gas Journal,* Jan. 28, 1985, pp. 119–24.
25. "Methanol Fuel Gets a Boost," *Chemical Week,* January 30, 1985, pp. 22–23.
26. "Methanol Suppliers Targeting Europe," *European Chemical News,* December 24/31, 1984, pp. 12–13.
27. "Key Chemicals—Methanol," *Chemical and Engineering News,* February 4, 1985.
28. *Chemical Industry Yearbook,* Second Edition. Surrey, England: Industrial Press, 1984.

Formaldehyde

1. List, H. L. "Formaldehyde—A Status Report," *International Petrochemical Developments,* Vol. 2, No. 19, October 1, 1981.
2. German Patent 2,743,113.
3. "Key Chemicals—Formaldehyde," *Chemical and Engineering News,* February 4, 1985.
4. "Chemical Profiles," *Chemical Marketing Reporter,* September 26, 1983.
5. *1984 Directory of Chemical Producers—United States,* SRI International.
6. "Effects of Foam Insulation Ban Far Reaching," *Chemical and Engineering News,* March 9, 1982, pp. 34–37.
7. "Formaldehyde's Fight for Self-Preservation," *Chemical Business,* March 8, 1982, pp. 33–38.
8. Konno, T. "Trends for High Concentration Technology in Formalin Production," *Chemical Economy and Engineering Review,* Vol. 13, No. 11, November 1981, pp. 31–36.
9. Mann, R. S. and M. K. Dosi. "Partial Oxidation of Methane to Formaldehyde Over Halogen Modified Catalyst," *J. Chem. Tech. Biotechnology,* Vol. 29, 1979, pp. 467–470.
10. French Patent 2,403,985 (April 20, 1979) to Chemische Werke Huels.
11. "Petrochemical Handbook '83," *Hydrocarbon Processing,* November 1983.
12. *1984 Directory of Chemical Producers—Western Europe,* SRI International.
13. *Kirk-Othmer Encyclopedia of Chemical Technology,* Third Edition, Vol. 11. New York: Wiley, 1984, pp. 231–250.
14. *JCW Chemicals Guide 82/83.* Japan: The Chemical Daily Co. Ltd., March 1982.
15. *Chemical Industry Yearbook,* Second Edition. Surrey, England: Industrial Press, 1984.

Acetic Acid

1. List, H. L. "Acetic Acid—A Status Report," *International Petrochemical Developments,* Vol. 3, No. 18, September 15, 1982.
2. *1984 Directory of Chemical Producers—United States,* SRI International.
3. "Chemical Profile," *Chemical Marketing Reporter,* May 9, 1983.
4. "Petrochemical Handbook '83," *Hydrocarbon Processing,* November 1983.
5. *1984 Directory of Chemical Producers—Western Europe,* SRI International.
6. *JCW Chemicals Guide 82/83.* Japan: The Chemical Daily Co. Ltd., March 1982.
7. Brown, A. S. "Technology Shift Tightens Acetic Acid Market," *Chemical Business,* November 1984, pp. 11–16.
8. *Kirk-Othmer Encyclopedia of Chemical Technology,* Third Edition, Vol. 1. New York: Wiley, 1984, pp. 124–47.
9. *Chemical Industry Yearbook,* Second Edition. Surrey, England: Industrial Press, 1984.

Acetic Anhydride

1. List, H. L. "Acetic Anhydride—A Status Report," *International Petrochemical Developments,* Vol. 3, No. 5, March 1, 1982.
2. "Chemical Profile," *Chemical Marketing Reporter,* May 16, 1983.
3. *1984 Directory of Chemical Producers—United States,* SRI International.
4. "Eastman's Big Investment Drive Nears Peak," *Chemical and Engineering News,* November 29, 1982, pp. 9–12.
5. "Petrochemical Handbook '83," *Hydrocarbon Processing,* November 1983.
6. *1984 Directory of Chemical Producers—Western Europe,* SRI International.
7. *Kirk-Othmer Encyclopedia of Chemical Technology,* Third Edition, Vol. 1. New York: Wiley, 1984, pp. 151–61.
8. *JCW Chemicals Guide 82/83.* Japan: The Chemical Daily Co. Ltd., March 1982.
9. *Chemical Industry Yearbook,* Second Edition. Surrey, England: Industrial Press, 1984.

Methyl Tertiary Butyl Ether

1. List, H. L. "Methyl Tertiary Butyl Ether—A Status Report," *International Petrochemical Developments,* Vol. 3, No. 12, June 15, 1982.
2. Marceglia, G. and G. Oriani. "The MTBE Market: Potential MTBE Demand, MTBE Production Options," *Chemical Economy and Engineering Review,* Vol. 14, No. 4 (No. 157), April 1982, pp. 35–45.
3. Ibid. "The Snamprogetti/Anic MTBE Technology, "*Chemical Economy and Engineering Review,* Vol. 14, No. 5 (No. 158), May 1982, pp. 37–42.
4. Smith, L. A. and M. N. Huddleston. "New MTBE Design Now Commercial," *Hydrocarbon Processing,* March 1982, pp. 121–23.
5. "Additives Widen Refiners Options in Making Gasolines," *Chemical and Engineering News,* April 9, 1984, pp. 13–14.
6. "U.S., European Refiners Prepare for Lead Cuts," *Chemical Engineering,* March 7, 1983, pp. 49–53.
7. Unzelman, G. H. "Value of Oxygenates in Blending Octane Improvements in Motor Gasoline Are Appraised," *Oil and Gas Journal,* June 1, 1981, pp. 77–84.

8. *1984 Directory of Chemical Producers—United States,* SRI International.
9. *1984 Directory of Chemical Producers—Western Europe,* SRI International.
10. *Kirk-Othmer Encyclopedia of Chemical Technology,* Third Edition, Vol. 11. New York: Wiley, 1984, pp. 652–95.
11. Unzelman, G. H. "Problems Hinder Full Use of Oxygenates in Fuel," *Oil and Gas Journal,* July 2, 1984, pp. 59–65.
12. Bitar, L. S., E. A. Hazbun, and W. J. Piel. "MTBE Production and Economics," *Hydrocarbon Processing,* October 1984, pp. 63–66.

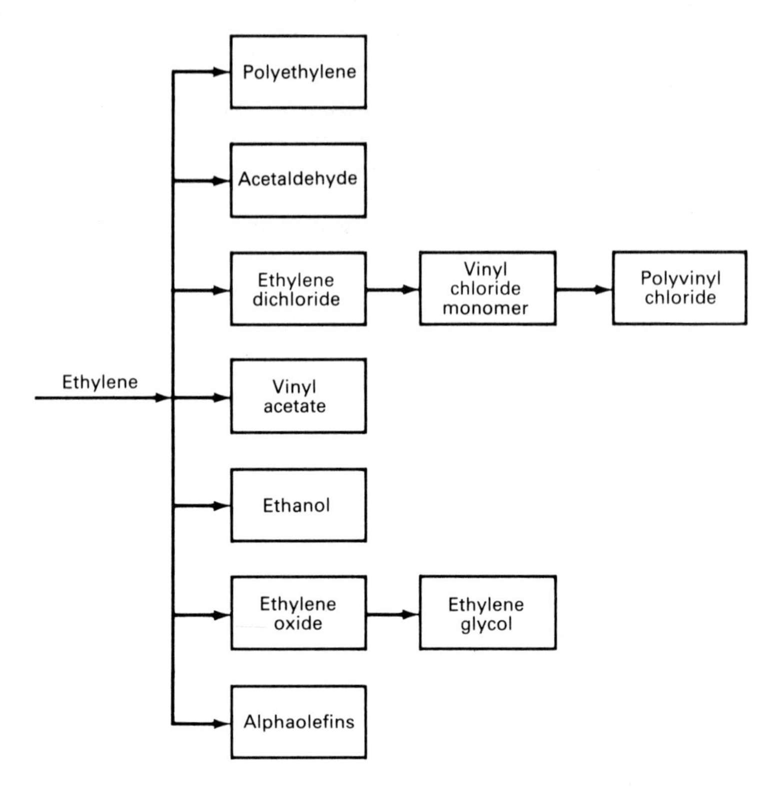

Ethylene derivatives

CHAPTER TWO

Ethylene Derivatives

ACETALDEHYDE

The primary sources of acetaldehyde in the United States and Europe are ethylene and ethanol, although some sizable facilities in Western Europe use acetylene as feedstock.

Technology

The Wacker process, developed by Hoechst of West Germany, involves the direct oxidation of ethylene using either air or oxygen. The catalyst is usually an aqueous solution of copper chloride and palladium chloride. The copper chloride acts as a promoter. The reactions are as follows:

$$C_2H_4 + PdCl_2 + H_2O \longrightarrow CH_3CHO + Pd + 2HCl$$

$$Pd + 2HCl + \tfrac{1}{2}O_2 \longrightarrow PdCl_2 + H_2O$$

The acetaldehyde is vaporized from the solution and absorbed in water, separated from the residual gases, and then distilled. Yields are approximately 95 percent. A block flowsheet for this process is shown in Figure 2-1.

With ethanol used as the feedstock, the feed is oxidized in the vapor state to acetaldehyde by passing it through a copper or silver gauze catalyst in accordance with the following equation:

$$C_2H_5OH + \tfrac{1}{2}O_2 \longrightarrow CH_3CHO + H_2O$$

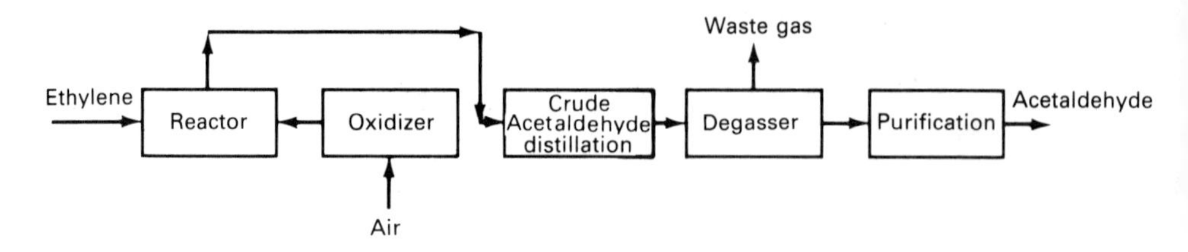

Figure 2.1 Production of acetaldehyde from ethylene (two-stage air process)

Yields are about 95 percent. The product is recovered by scrubbing with ethanol; pure product is recovered by distillation.

With acetylene used as a feedstock, the raw material is passed into a suspension of ferric and mercurous sulfates in sulfuric acid under a pressure of about 1 atmosphere and a temperature of about 100°C. Yields are about 95 percent, and the reaction is:

$$C_2H_2 + H_2O \longrightarrow CH_3CHO$$

Market

Major end uses for acetaldehyde are acetic acid and acetic anhydride, which account for perhaps half of the demand. All new acetic acid plants are based on methanol, which is expected to phase out this use for acetaldehyde in a few years.

Currently only two producers of acetaldehyde are in the United States—Celanese and Eastman. Celanese uses ethylene feedstock, and Eastman uses mostly ethylene with a small amount of ethanol. Their capacities are shown in Table 2-1 (1).

In addition to the facilities listed in Table 2-1, Celanese has a 450 million pound per year facility at Clear Lake, Texas, which has been on standby since 1981.

Major Western European producers, capacities, and feedstocks are shown in Table 2-2 (2).

TABLE 2-1 Major U.S. Producers of Acetaldehyde

		Estimated Capacity	
Producer	Location	mm lb/yr	m MT/yr
Celanese	Bay City, Texas	250	113
Eastman	Longview, Texas	500	227
	TOTAL	750	340

TABLE 2-2 Major Western European Producers of Acetaldehyde

Producer	Location	Estimated Capacity mm lb/yr	m MT/yr	Feedstock
Hoechst	Lillebonne, France	212	96	Ethylene
Chem. Werke Huels	Marl, W. Germany	265	120	Acetylene
	Wanne-Eickel, W. Germany	176	80	Ethanol
Hoechst	Frankfurt, W. Germany	375	170	Ethylene
	Hurth-Knapsack, W. Germany	293	133	Ethylene
Wacker Chemie	Burghausen, W. Germany	132	60	Ethylene
Montedison	Priolo, Italy	397	180	Ethylene
Ind. Quim. Asoc.	Tarragona, Spain	298	135	Ethylene
Union Exp. RioTinto	Guardo, Spain	44	20	Acetylene
Lonza	Visa, Switzerland	88	40	Ethylene
	TOTAL	2280	1034	

Japan has about 11 major producers of acetaldehyde, including Chisso Petrochemical, Japan Aldehyde, Mitsubishi Chemical Industries, Mitsui Petrochemical Industries, Showa Denko, Sumitomo Chemical, and Tokuyama Petrochemical.

Economics

Typical production costs for the Wacker two-stage process using air are as follows:

Capacity: 110 million lb/yr (50 m MT/yr)
Capital cost: BLCC, \$19 mm; OSBL, \$8 mm; WC, \$8 mm

	¢/lb	\$/MT	%
Raw materials[a]	15.0	332	67
Utilities	2.2	49	10
Operating costs	1.1	24	5
Overhead costs	4.1	90	18
Cost of production	22.4	495	100
Transfer price	7.4	163	

[a] Ethylene at 22¢/lb

ETHYLENE DICHLORIDE AND VINYL CHLORIDE MONOMER

Vinyl chloride monomer (VCM) is largely used in the manufacture of polyvinyl chloride and will obviously follow the market for this polymer. PVC production, worldwide, is exceeded in the production of polymers by only polyethylene.

VCM was originally manufactured by reacting acetylene with hydrogen chloride, but this route has been largely replaced by ethylene-based technology. In the mid-seventies VCM was shown to be carcinogenic, which resulted in modification in the manufacturing technology of both the monomer and PVC.

Technology

Although some VCM is still manufactured from acetylene, any new capacities are not expected to be based on this feedstock. The acetylene-based technology was replaced by the direct chlorination of ethylene to produce ethylene dichloride, which is then pyrolyzed to VCM and HCl.

$$C_2H_4 + Cl_2 \longrightarrow CH_2ClCH_2Cl$$

$$CH_2ClCH_2Cl \longrightarrow CH_2{=}CHCl + HCl$$

The chlorination is usually carried out in the liquid phase using ethylene dichloride as a solvent. Conditions are moderate—about 4 atmospheres pressure and 60°C. Ferric chloride catalysts are used and the reaction is rapid, with conversions of ethylene and selectivity to the ethylene dichloride both approximately 98 percent. The pyrolysis of the ethylene dichloride is carried out at about 400 psia and 555°C, with conversions of about 60 percent and selectivities of about 95 percent.

By the late fifties a process involving the oxychlorination of ethylene was commercialized. In this technology ethylene is reacted with oxygen and hydrogen chloride to produce ethylene dichloride and water as follows:

$$C_2H_4 + 2HCl + \tfrac{1}{2}O_2 \longrightarrow CH_2ClCH_2Cl + H_2O$$

The reaction is carried out at a temperature of about 275°C and at pressures of up to 200 psia. With supported metal chloride catalysts, conversions and selectivity are quite high. The pyrolysis of the ethylene dichloride proceeds as described previously. Presently in the United States the direct chlorination, oxychlorination, and ethylene dichloride reactions are often balanced so that only vinyl chloride and water are produced, thereby eliminating any by-product hydrogen chloride, which can be a problem. A block flow diagram of a balanced facility is shown in Figure 2-2.

In a balanced world-scale facility, raw materials, primarily ethylene and chlorine, account for about three-fourths of the cost of production, with capital investment accounting for most of the remaining cost. Utilities and operating costs are generally quite minor.

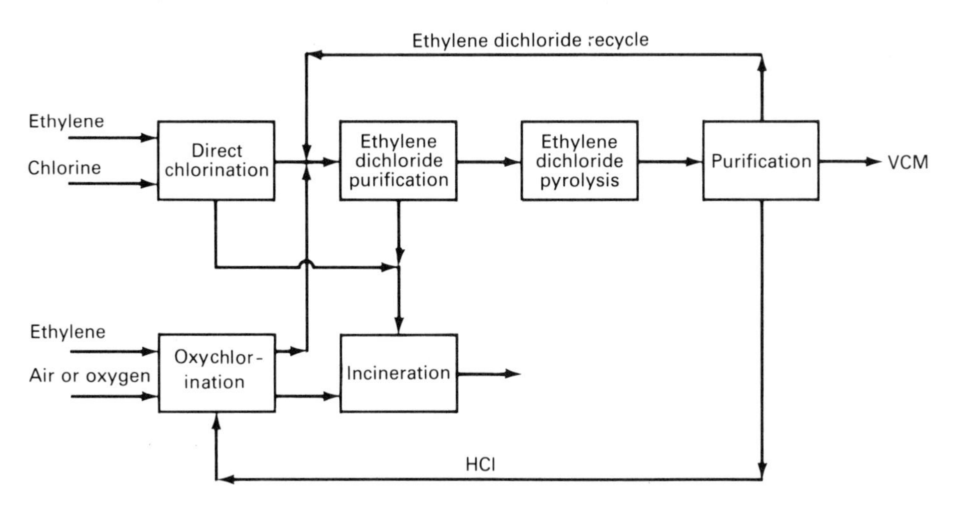

Figure 2.2 Production of vinyl chloride monomer—balanced oxychlorination process

Market

As mentioned earlier, virtually all of the vinyl chloride monomer is used for vinyl polymers and, subsequently, fabricated vinyl plastic items. Since VCM was shown to be carcinogenic, care must be taken to sharply reduce the VCM content of the finished polymer. Further details will be discussed in the next section, on PVC.

Worldwide, nameplate capacity of VCM is in excess of 25 billion pounds per year, of which over 9 billion pounds are in facilities in the United States. Producers in the United States of ethylene dichloride are shown in Table 2-3 (3). Carbide's production is for captive use only. Shell closed a 1.2 billion pound per year facility at Norco, Louisiana, for an indefinite period because of a temporary closing of its 700 million pound per year VCM facility at the same location. Production of the material in 1983 was in excess of 10 billion pounds, with 84 percent going to VCM and 9 percent being exported. The remainder was used for chlorinated solvents and miscellaneous applications. Projections for 1987 are in excess of 12 billion pounds, and exports are expected to grow slightly (3). Major Western European producers of ethylene dichloride are shown in Table 2-4 (6).

U.S. production of VCM is shown in Table 2-5 (4). Dow is the largest merchant producer of VCM in the United States. Western Europe has approximately 30 producers of VCM, with nameplate capacity of about 13 billion pounds per year. Details are shown in Table 2-6 (6).

TABLE 2-3 Major U.S. Producers of Ethylene Dichloride

Producer	Location	Estimated Capacity	
		mm lb/yr	m MT/yr
Arco	Port Arthur, Texas	500	226
Borden	Geismar, Louisiana	510	231
Diamond Shamrock	Deer Park, Texas	190	86
Dow	Freeport, Texas	1600	726
	Oyster Creek, Texas	1200	544
	Plaquemine, Louisiana	2000	907
Ethyl	Pasadena, Texas	225	102
	Baton Rouge, Louisiana	700	317
Formosa Plastic	Baton Rouge, Louisiana	550	249
	Point Comfort, Texas	850	385
Georgia-Pacific	Plaquemine, Louisiana	1650	748
B. F. Goodrich	LaPorte, Texas	1585	719
	Calvert City, Kentucky	1000	454
	Convent, Louisiana	800	363
PPG	Lake Charles, Louisiana	2700	1224
Shell	Deer Park, Texas	1400	635
Union Carbide	Taft, Louisiana	150	68
	Texas City, Texas	150	68
Vulcan	Geismar, Louisiana	350	159
	TOTAL	18,110	8211

TABLE 2-4 Major Western European Producers of Ethylene Dichloride

Producer	Location	Estimated Capacity	
		mm lb/yr	m MT/yr
BASF	Antwerp, Belgium	419	190
Soc. Limbourg. duVin.	Tessenderlo, Belgium	1588	720
Solvic	Jemeppe-sur-Sambre, Belgium	882	400
ATOCHEM	Lavera, France	1455	660
	Saint-Auban, France	397	180
Soc. duChlor. deVin.	Fos-sur-Mer, France	728	330
Solvey and Cie	Tavaux, France	882	400
BASF	Ludwigshafen, W. Germany	364	165
Chem. Werke Huels	Marl, W. Germany	529	240
Deutsch Solvay-Werke	Rheinberg, W. Germany	772	350
Dow Chemical	Stade, W. Germany	595	270
Dynamit Nobel	Niederkasse, W. Germany	397	180
Hoechst	Gendorf, W. Germany	589	267
	Hurth-Knapsack, W. Germany	287	130

TABLE 2-4 Major Western European Producers of Ethylene Dichloride (*continued*)

Producer	Location	Estimated Capacity mm lb/yr	m MT/yr
ICI	Wilhelmshaven, W. Germany	1103	500
Wacker-Chemie	Burghausen, W. Germany	860	390
Ethyl Hellas Chem.	Thessaloniki, Greece	110	50
Enichem	Assemini, Italy	221	100
	Gela, Italy	485	220
	Porto-Torres, Italy	551	250
	Priolo, Italy	386	175
Montedison	Porto-Marghera, Italy	706	320
Rivede Sri	Brindisi, Italy	750	340
Rotterdamse Vinyl.	Rotterdam, Netherlands	1852	840
Norsk Hydro	Rafnes, Norway	1103	500
	TOTAL	18,011	8167

Nameplate capacity of VCM in Japan is about 5 billion pounds per year. The largest producers are Kanegafuchi, Kashima VCM, and Ryo-Nichi. Integration between VCM and PVC producers is more prevalent than in the United States. More than three-fourths of the VCM production is by companies that are also PVC producers. Major Japanese producers of EDC and VCM are shown in Table 2-7.

TABLE 2-5 U.S. Producers of VCM

Producer	Location	Estimated Capacity mm lb/yr	m MT/yr
Borden	Geismar, Louisiana	610	277
Dow	Oyster Creek, Texas	750	340
	Plaquemine, Louisiana	1250	567
DuPont	Lake Charles, Louisiana	700	317
Formosa Plastics	Baton Rouge, Louisiana	300	136
	Point Comfort, Texas	530	240
Georgia-Pacific	Plaquemine, Louisiana	1000	454
B. F. Goodrich	Calvert City, Kentucky	1000	454
	LaPorte, Texas	1000	454
PPG	Lake Charles, Louisiana	900	408
Shell	Deer Park, Texas	840	381
	Norco, Louisiana	700	317
	TOTAL	9580	4345

TABLE 2-6 Western European Producers of VCM

Producer	Location	Estimated Capacity	
		mm lb/yr	m MT/yr
BASF	Antwerp, Belgium	331	150
Soc. Limbourg. duVinyl.	Tessenderlo, Belgium	992	450
Solvic	Jemeppe-Sur-Sambre, Belgium	485	220
ATOCHEM	Lavera, France	882	400
	Saint-Auban, France	243	110
Soc. Chlor. deVinyl.	Fos-Sur-Mer, France	441	200
Solvic	Tavaux, France	529	240
BASF	Ludwigshafen, W. Germany	331	150
Chem. Werke Huels	Marl, W. Germany	904	410
Deutsche Solvay-Werke	Rheinberg, W. Germany	518	235
Hoechst	Gendorf, W. Germany	353	160
	Hurth-Knapsack, W. Germany	221	100
ICI	Wilhelmshaven, W. Germany	662	300
Wacker-Chemie	Burghausen, W. Germany	496	225
Ethyl Hellas Chemical	Thessaloniki, Greece	49	22
Enichem	Assemini, Italy	198	90
	Brindisi, Italy	420	190
	Porto-Marghera, Italy	639	290
	Porto-Torres, Italy	331	150
	Ravenna, Italy	176	80
Rotterdamse Vinyl.	Rotterdam, Netherlands	1103	500
Norsk Hydro	Rafnes, Norway	662	300
SARL-CIRES	Estarreja, Portugal	33	15
Aiscondel	Vilaseca, Spain	331	150
Viniclor	Martorell, Spain	485	220
Nordisk Philblack	Stenungsund, Sweden	221	100
ICI	Fleetwood, U.K.	441	200
	Runcorn, U.K.	441	200
	TOTAL	12,918	5857

Japan is expected to become a major importer of EDC and VCM when they close down some of their mercury cell chlor-alkali capacity because of environmental problems. Shell will start up a 1 billion pound per year facility at al-Jubail in Saudi Arabia in the mid-eighties, and much of the output is expected to go to Japan.

The concern regarding carcinogenicity of VCM was mentioned previously. Present emission standards are 1 ppm based on evidence of liver cancer caused by vinyl chloride. Standards could possibly be tightened in the future, and the result could be a further modification in manufacturing technology.

TABLE 2-7 Major Japanese Producers of EDC and VCM

		Estimated Capacity, EDC		Estimated Capacity, VCM	
Producer	Location	mm lb/yr	m MT/yr	mm lb/yr	m MT/yr
Denki Kagaku	Omi	—	—	93	42
Kanegafuchi	Takasago	920	417	550	249
Kureha	Nishiki	185	84	110	50
Mitsui Toatsu	Nagoya	243	110	146	66
Mitsui Senpoku	Sakai	243	110	146	66
Mitsubishi-Monsanto	Yokkaichi	295	134	176	80
Sumitomo	Kikumoto	203	94	121	55
Toa Gosei Chemical	Takushima	180	82	108	49
Asahi-Penn Chemical	Goi	203	94	121	55
Toyo Soda	Yokkaichi	467	212	280	127
	Nanyo	633	287	380	172
Nissan Chem. Ind.	Goi	220	100	132	60
Sun Arrow	Tokuyama	440	200	285	129
Ryo-Nichi	Mizushima	810	367	485	220
Sanyo Monomer	Mizushima	550	249	330	150
Kashima VCM	Mizushima	995	451	595	270
Chiba VCM	Sodegaura	290	132	350	159
Central Glass	Kawasaki	145	66	175	79
	TOTAL	7022	3189	4583	2078

Economics

Typical production costs for direct chlorination of ethylene to ethylene dichloride and a balanced oxychlorination process are as follows:

Ethylene Dichloride

Capacity: 1.7 billion lb/yr (770 m MT/yr)
Capital cost: BLCC, \$21 mm; OSBL, \$12 mm; WC, \$64 mm

	¢/lb	\$/MT	%
Raw materials[a]	8.6	190	76
Utilities	1.9	42	17
Operating costs	0.1	2	1
Overhead costs	0.7	15	6
Cost of production	11.3	249	100
Transfer price	11.9	262	

[a] Ethylene at 22¢/lb.; chlorine at 6¢/lb

Vinyl Chloride Monomer

Capacity: 1.0 billion lb/yr (454 m MT/yr)
Capital cost: BLCC, $125 mm; OSBL, $71 mm; WC $68 mm

	¢/lb	$/MT	%
Raw materials[a]	15.3	337	75
Utilities	1.4	31	7
Operating costs	0.8	18	4
Overhead costs	2.8	62	14
Cost of production	20.3	448	100
Transfer price	26.2	578	

[a] Ethylene at 22¢/lb; chlorine at 6¢/lb

POLYVINYL CHLORIDE

Most of the vinyl chloride monomer is used in the production of polyvinyl chloride (PVC). PVC is second in the world in polymer production, exceeded only by polyethylene.

Technology

PVC is manufactured commercially by the polymerization of vinyl chloride initiated by a free radical. The major techniques are suspension polymerization, emulsion polymerization, and mass (or bulk) polymerization. A fourth technique, solution polymerization, is used primarily in the production of copolymers of vinyl chloride and vinyl acetate. In this technique the polymerization occurs in the presence of an organic diluent. The system is free of emulsifiers or surface-active agents. The polymers are, with regard to purity, similar to those obtained from mass polymerization.

Suspension polymerization. This technique is the most widely used for PVC. The monomer is finely dispersed in water by means of agitation. An initiator, soluble in VCM, begins the polymerization reaction. In addition, a suspension stabilizer is usually added. This stabilizer inhibits coalescence of the droplets of monomer and, by forming a layer on the PVC particles, prevents their agglomeration. Most of the improvements in this process have been directed toward the initiator and the suspension system. The most common initiation mechanism is from the decomposition of peroxides. Decomposition of azo compounds is also used. In some cases combinations of initiators are employed. The initiator and its concentration control the rate of the polymerization reaction. Higher operating temperatures tend to promote lower-molecular-weight polymers. A block flow diagram for this process is shown in Figure 2-3.

The suspension system controls the particle size and the polymer porosity. Common suspension systems include methyl cellulose and polyvinyl alcohol as well as a surfactant to increase polymer porosity. The free radical polymerization of vinyl chloride goes through steps, including decomposition of the initiator, propagation of a reactive radical of the monomer, chain transfer, and finally chain termination.

Emulsion polymerization. This technique utilizes a soap or surface-active agent and a water-soluble initiator to produce resins that generally have small particle diameters and narrow particle size distributions. Unfortunately somewhat larger amounts of soap often remain with the polymer product and reduce its clarity, electrical resistance, and water resistance.

The emulsifier exists as an aggregate of colloidal molecules arranged with the hydrophilic ends at the aqueous phase and the hydrophobic ends within. These aggregates are called micelles. Monomer from the emulsion droplets in the aqueous phase diffuses through the emulsifier molecule wall and into the micelle. The initiator produces a free radical, and polymerization proceeds through initiation, propagation, chain transfer, and chain termination in a manner similar to suspension polymerization. In general, emulsion polymerization produces resins of somewhat higher molecular weight than suspension polymerization does. Initiators, which must be water soluble, are typically hydrogen peroxide, potassium persulfate, and ammonium persulfate. Emulsifiers include ammonium stearate, sodium lauryl sulfate, and sulfonated higher alkyl alcohols.

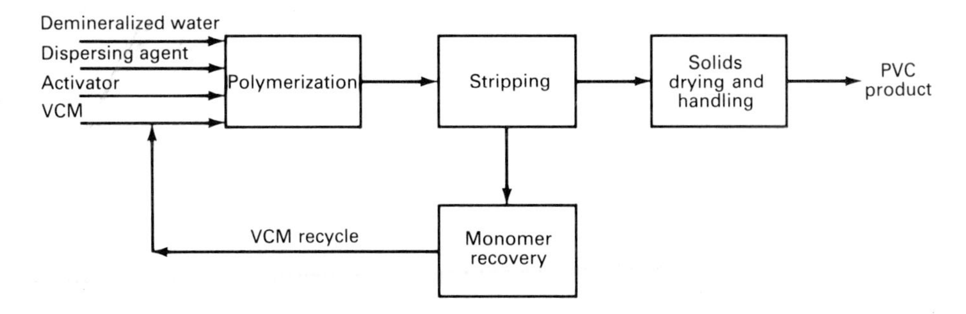

Figure 2.3 Production of PVC—suspension process

Mass (or bulk) polymerization. Mass polymerization proceeds without water or solvent. Since only a monomer soluble-free radical initiator is used, the resins produced are free of soap or surfactant residues and have high porosity and clarity. The particles generally range from 0.5 to 1 micron, are uniform in shape, and have a narrow particle size distribution. They are particularly useful for those applications that require strength, transparency, high electrical resistance, and heat stability.

The polymerization begins with decomposition of free radical initiator, which is similar to the initiator used in suspension polymerization. As polymerization proceeds, PVC particles are precipitated from the remaining monomer, since they are insoluble. As the polymer precipitates, agglomerates form and absorb monomer. At high conversions the heterogeneous reaction mixture becomes viscous and causes local overheating problems. These localized temperature variations result in a lower-molecular-weight product and a wider-molecular-weight distribution. Improvements in this technique are primarily directed toward avoiding these problems. Commercial mass polymerization reactions are generally carried out at about 60°C and 10 atmospheres.

The most dramatic technological change in recent years has been the change from reactors having a capacity of only 1000 gallons to capacities of up to about 36,000 gallons and even larger. The Japanese have pioneered the use of these large reactors.

Market

PVC became commercially important about 40 years ago, replacing rubber in many applications. The growth has been remarkable, and PVC is second only to polyethylene in world polymer production.

U.S. production of PVC and copolymers went from about 3.6 billion pounds in 1975 to about 6.2 billion pounds in 1979 and about 5.5 billion pounds in 1983. In 1979 Western European production was almost 9 billion pounds and Japanese production was over 3 billion pounds. The worldwide recession, particularly the decrease in construction, was responsible for the reduction of output in the early eighties.

Major U.S. producers of PVC and estimated capacities of their installations are shown in Table 2-8 (2,13). Capacities shown in Table 2-8 vary with product mix. Union Carbide produces only specialty grades of PVC. Several smaller facilities, such as Talleyrand's facility in Massachusetts and Occidental's facility in Maryland, are idled. The figures in Table 2-8 show that a significant overcapacity is available despite the numerous consolidations that have already occurred. The need for additional capacity is not expected for many years.

TABLE 2-8 Major U.S. Producers of PVC

Producer	Location	Estimated Capacity	
		mm lb/yr	m MT/yr
Air Products	Pensacola, Florida	200	91
	Calvert City, Kentucky	200	91
Borden	Illiopolis, Illinois	340	154
	Leominster, Massachusetts	185	84
	Geismer, Louisiana	300	136
Certain-Teed	Lake Charles, Louisiana	225	102
Conoco (DuPont)	Aberdeen, Mississippi	455	206
	Oklahoma City, Oklahoma	260	118
Formosa Plastics	Delaware City, Delaware	310	141
	Point Comfort, Texas	540	245
General Tire	Ashtabula, Ohio	120	54
Georgia-Pacific	Plaquemine, Louisiana	700	317
	Delaware City, Delaware	150	68
B. F. Goodrich	Avon Lake, Ohio	400	181
	Deer Park, Texas	260	118
	Henry, Illinois	200	91
	Long Beach, California	150	68
	Louisville, Kentucky	375	170
	Pedricktown, New Jersey	400	181
	Plaquemine, Louisiana	190	86
Goodyear	Niagara Falls, New York	70	32
Keysor	Saugus, California	50	23
Occidental	Baton Rouge, Louisiana	230	104
	Burlington, New Jersey	180	82
	Pottstown, Pennsylvania	180	82
Pantasote	Passaic, New Jersey	50	23
Shintech	Freeport, Texas	660	299
Tenneco	Burlington, New Jersey	160	73
	Flemington, New Jersey	100	45
	Pasadena, Texas	700	317
Union Carbide	South Charleston, W. Virginia	50	23
	Texas City, Texas	100	45
	TOTAL	8490	3850

Major applications of PVC and the percentages used for the different U.S. markets are shown in Table 2-9. PVC is a mature industrial material, and the only changes taking place are in its markets. The market is about evenly divided between rigid and flexible material.

Major Western European producers of PVC are listed in Table 2-10 (11). Major Japanese producers of PVC are shown in Table 2-11.

TABLE 2-9 Major Applications of PVC for U.S. Markets

Application	% Used
Rigid pipe and tubing, including fittings	40.0
Flooring, textile, etc.	11.1
Coatings and pastes	10.8
Wire and cable	6.6
Film and sheet	4.5
Miscellaneous extrusion uses	7.8
Phonograph records	2.6
Bottles	1.3
Miscellaneous molding uses	2.4
Exports	5.1
Other	7.8

TABLE 2-10 Major Western European Producers of PVC

		Estimated Capacity	
Producer	Location	mm lb/yr	m MT/yr
Halvic Kunststoffwerke	Hallein, Austria	132	60
BASF	Antwerp, Belgium	221	100
Solvic	Jemeppe-Sur-Sambre, Belgium	496	225
Neste Oy	Kulloo, Finland	132	60
ATOCHEM	Balan, France	276	125
	Brignoud, France	276	125
	Saint-Auban, France	143	65
	Saint-Fons, France	397	180
Shell Chimie	Berre-L'Etang, France	320	145
Soc. Artesienne de Vin.	Mazingarbe, France	353	160
Solvic	Tavaux, France	551	250
BASF	Ludwigshafen, W. Germany	331	150
Chemische Werke Huels	Marl, W. Germany	904	410
Deutsche ICI	Waldshut, W. Germany	71	32
Deutsche Solvay-Werke	Rheinberg, W. Germany	364	165
Hoechst	Gendorf, W. Germany	276	125
ICI	Wilhelmshafen, W. Germany	254	115
Wacker-Chemie	Burghausen, W. Germany	396	180
	Koln, W. Germany	375	170
Esso Pappas Chemical	Thessaloniki, Greece	104	47
Enichem Polimeri	Assemini, Italy	254	115
	Brindisi, Italy	397	180
	Porto-Marghera, Italy	706	320
	Porto-Torres, Italy	243	110
	Ravenna, Italy	309	140
Enochimica del Montfer.	Terni, Italy	88	40
Solvic	Ferrara, Italy	265	120
DSM	Geleen, Netherlands	396	180

TABLE 2-10 Major Western European Producers of PVC (*continued*)

Producer	Location	Estimated Capacity	
		mm lb/yr	m MT/yr
Rotterdamse Vinylunie	Rotterdam, Netherlands	396	180
Norsk Hydro	Porsgrunn, Norway	154	70
SARL-CIRES	Estarreja, Portugal	132	60
Aiscondel	Monzon-Del-Rio-Cinca, Spain	110	50
	Vilaseca, Spain	88	40
Hispavic Industrial	Martorell, Spain	198	90
	Torrelavega, Spain	44	20
Rio Rodano	Hernani, Spain	66	30
	Miranda-de-Ebro, Spain	176	80
Nordisk Philblack	Stenungsund, Sweden	221	100
	Sundsvall, Sweden	44	20
ICI	Sins, Switzerland	77	35
	Barry, Wales, U.K.	198	90
	Hillhouse, U.K.	396	180
	Runcorn, U.K.	243	110
Norsk Hydro Polymers	Newton Aycliffe, U.K.	198	90
	Stavely, U.K.	55	25
	TOTAL	11,826	5364

TABLE 2-11 Major Japanese Producers of PVC

Producer	Location	Estimated Capacity	
		mm lb/yr	m MT/yr
Asahi Glass	Chiba	62	28
Central Glass	Kawasaki	60	27
Chisso	Minamata, Goi, Mizushima	320	145
Denki Kagaku Kogyo	Omi, Chiba, Shibukawa	390	177
Kanagafuchi Chem. Ind.	Osaka, Takasago, Kashima	475	215
Kureha Chem. Ind.	Nishiki	300	136
Mitsubishi Monsanto	Yokkaichi	270	122
Mitsui Toatsu	Sakai, Nagoya	255	116
Nippon Zeon	Takaoka, Mizushima	420	190
Nissan Chem. Ind.	Chiba	60	27
Ryo-Nichi	Uozu, Mizushima	300	136
Shin-Etsu Chemical	Tokuyama, Kashima	570	259
Sumitomo Chemical	Kikimoto, Chiba	255	116
Sun Arrow	Tokuyama	130	59
Toagosei Chem. Ind.	Tokushima	190	86
Tokuyama Sekisui	Tokuyama	110	50
Toyo Soda	Yokkaichi	150	68
	TOTAL	4317	1957

Economics

Typical production costs for a suspension process facility are as follows:

Capacity: 300 mm lb/yr (136 m MT/yr)
Capital cost: BLCC, $45 mm; OSBL, $18 mm; WC, $35 mm

	¢/lb	$/MT	%
Raw materials[a]	29.0	639	82
Utilities	1.6	35	5
Operating costs	1.0	22	3
Overhead costs	3.8	84	10
Cost of production	35.4	780	100
Transfer price	41.7	919	

[a] Vinyl chloride monomer at 28¢/lb

VINYL ACETATE

The market for vinyl acetate in the early eighties was severely depressed in the United States and worldwide primarily because of the depressed state of the housing industry. Most of the vinyl acetate is used in end products such as adhesives and paints, which are largely dependent on housing construction. Other substantial outlets are in the production of polyvinyl alcohol and as a copolymer with other olefinic monomers. Most processes are based on ethylene, but a few still begin with acetylene. Research continues for processes that begin with synthesis gas.

Technology

Ethylene-based processes. Both liquid-phase and vapor-phase ethylene-based processes were developed in the sixties. Several companies in the United States, Europe, and Japan were involved in the liquid-phase technology, but only the ICI process has survived. Severe corrosion problems were the major reason for the discontinuation. The vapor-phase process was developed primarily in the United States by U.S. Industrial Chemicals and in Germany by Bayer and Hoechst. The process involves a noble-metal-catalyzed oxidation, with the overall reaction as follows:

$$CH_2{=}CH_2 + CH_3OH + \tfrac{1}{2}O_2 \longrightarrow CH_2{=}CHOOCCH_3 + H_2O$$

In general, operating pressures are about 4 to 8 atmospheres and temperatures about 150 to 180°C. Selectivities to vinyl acetate are better than 90 mole percent based on ethylene. Much of the unconverted 10 mole percent involves the production of carbon dioxide, with some acetaldehyde and ethyl acetate also being produced. Pure oxygen is used rather than air, because recovery of unreacted ethylene would require the elimination of large quantities of nitrogen. The key element in the catalyst system is palladium. After the reaction the product recovery involves the following steps:

1. Quench: condensing out most of the acetic acid, water, and vinyl acetate.
2. Recovery: scrubbing with acetic acid to recover any vinyl acetate and then scrubbing with water to recover acetic acid.
3. Carbon dioxide removal: typically using a hot carbonate system.
4. Product purification: distilling the product liquids to separate vinyl acetate from water and impurities.

A block flow diagram for the process is shown in Figure 2-4.

Acetylene-based process. The acetylene-based process was developed in Germany in the thirties, and some installations still exist for the vapor-phase reaction of acetylene with acetic acid using a carbon-supported zinc acetate catalyst. The generally higher cost for acetylene feedstock has largely resulted in the supplantation of this process by the ethylene-based process.

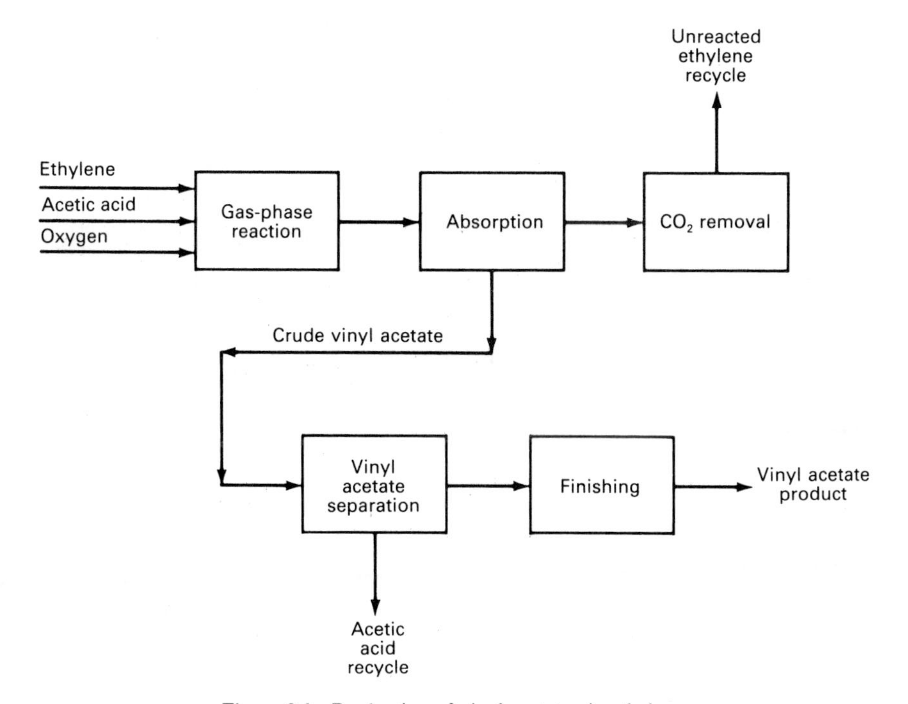

Figure 2.4 Production of vinyl acetate via ethylene

Synthesis-gas-based processes. Many companies worldwide have been actively pursuing a series of processes to produce acetyl compounds via carbonylation reactions employing synthesis gas. In the United States Halcon has been very active (2). A facility for the production of acetic anhydride by carbonylation technology has been built by Tennessee-Eastman at Kingsport, Tennessee, using coal as the basic feedstock and is being watched carefully by those ultimately considering a coal-based chemicals business. A process for the production of vinyl acetate would probably involve the following steps:

1. Reaction of methanol with recycled acetic acid to produce methyl acetate.
2. Carbonylation and hydrogenation of the methyl acetate with synthesis gas to produce ethylidene diacetate.
3. Pyrolysis of the ethylidene diacetate to produce vinyl acetate and acetic acid for recycle.

Note that the starting point, methanol, can be produced from synthesis gas. The catalyst system is understood to consist of a rhodium salt modified by 3-picoline and promoted by methyl iodide.

Ultimately all the processes available or being developed will be tied to the price and availability of ethylene, acetylene, or synthesis gas.

Market

U.S. demand for vinyl acetate in 1983 was close to 2 billion pounds and is projected to grow to about 2.3 billion pounds in 1987 (3). The largest end uses, polyvinyl acetate emulsions and resins, are projected to grow by about 5 percent through the rest of this decade. Exports from the United States will undoubtedly drop from their current level of almost 30 percent, because of a decrease in feedstock advantage as well as the start-up of new facilities in such countries as Canada, Mexico, Brazil, and Taiwan. End uses in the United States are shown in the following table (1):

End Uses of Vinyl Acetate in the United States

End Uses	% Used
Polyvinyl acetate emulsions and resins	40
Polyvinyl alcohol	16
Polyvinyl butyral and formal	5
Ethylene-vinyl acetate copolymers	5
Polyvinyl chloride copolymers	4
Exports	30

The major producers of vinyl acetate are shown in Table 2-12 for the United States (3,6,7), Table 2-13 for Western Europe (5), and Table 2-14 for Japan.

TABLE 2-12 Major U.S. Producers of Vinyl Acetate

Producer	Location	Estimated Capacity	
		mm lb/yr	m MT/yr
Borden	Geismer, Louisiana	115 (shut down)	52
Celanese	Bay City, Texas	425	193
	Clear Lake, Texas	425	193
DuPont	LaPorte, Texas	400	181
Union Carbide	Texas City, Texas	550	249
USI	Deer Park, Texas	600	272
	TOTAL	2515	1140

TABLE 2-13 Major Western European Producers of Vinyl Acetate

Producer	Location	Estimated Capacity		Feedstock
		mm lb/yr	m MT/yr	
Rhone-Poulenc	Pardies, France	309	140	Acetylene
Hoechst	Frankfort, W. Germany	331	150	Ethylene
Wacker-Chemie	Burghausen, W. Germany	176	80	Ethylene
Montedison	Porto-Marghera, Italy	121	55	
Monomeros Espan.	Guardo, Spain	55	25	Acetylene
BP Chemicals	Port Talbot, Wales, U.K.	176	80	Ethylene
	TOTAL	1168	530	

TABLE 2-14 Major Japanese Producers of Vinyl Acetate

Producer	Location	Estimated Capacity		Feedstock
		mm lb/yr	m MT/yr	
Kuraray	Nakajo	190	86	Acetylene
	Okayama	250	113	Ethylene
Nippon Synth. Chem.	Mizushinma	290	132	Ethylene
Denki Kagaku Kogyo	Chiba	130	59	Ethylene
	Omi	130	59	Acetylene
Shin-Etsu	Sakai	175	79	Ethylene
Showa Denko	Ohita	175	79	Ethylene
	TOTAL	1340	607	

Economics

Typical production costs for processes beginning with ethylene and with acetylene are as follows:

From Ethylene

Capacity: 350 mm lb/yr (159 m MT/yr)
Capital cost: BLCC, $73 mm; OSBL, $30 mm; WC, $42 mm

	¢/lb	$/MT	%
Raw materials[a]	27.0	595	76
Utilities	3.0	66	8
Operating costs	1.2	26	3
Overhead costs	4.6	101	13
Cost of production	35.8	789	100
Transfer price	44.6	983	

[a] Acetic acid at 25¢/lb; ethylene at 22¢/lb

From Acetylene

Capacity: 250 mm lb/yr (113 m MT/yr)
Capital cost: BLCC, $37 mm; OSBL, $22 mm; WC, $33 mm

	¢/lb	$/MT	%
Raw materials[a]	31.2	688	79
Utilities	3.0	66	8
Operating costs	1.0	22	3
Overhead costs	4.0	88	10
Cost of production	39.2	864	100
Transfer price	46.3	1021	

[a] Acetic acid at 25¢/lb; acetylene at 40¢/lb

ETHANOL

The past decade has shown substantial alterations in the worldwide ethanol business. Ten years ago ethanol was produced in virtually all areas of the world. It was used primarily for beverage, solvent, and chemical applications, but the increasing energy prices and concern over the reliability of supplies resulted in attention being paid to the use of ethanol as a fuel extender and, in the pure state, as a fuel. Since the material can be produced from a wide variety of feedstocks by fermentation, energy-short countries with agricultural resources began considering ethanol as a fuel to reduce their dependence on imported oil. The past few years, however, have seen a stabilization of crude oil prices, and fuel ethanol programs in many parts of the world were postponed or canceled. However, many facilities were constructed, which will promote substantial growth in the worldwide ethanol business.

Technology

Most industrial fermentation involves the production of ethyl alcohol from carbohydrates via yeast. Some sugar-containing materials, such as sugarcane juice and molasses, can be directly fermented in this manner, but other natural carbohydrates, such as starch and cellulose, must first be hydrated to fermentable hydrates. This process is often accomplished by the action of enzymes. The major feedstock for the production of fermentation ethanol in the United States is cornstarch; in most other parts of the world, sugarcane molasses is used. The last decade has seen substantial growth in the fermentation ethanol industry, primarily because of government subsidies. The catalyst for this growth was ethanol's potential use in the motor fuel market, but it is also impinging on the industrial market. Since ethanol for industrial applications is predominately supplied from synthetic ethanol, which is produced by ethylene hydration, production cost comparisons for synthetic versus fermentation ethanol is probably the key to projecting the future of synthetic and fermentation alcohol.

The formation of sugars from the starch in the corn involves hydration, gelatinization, and hydrolysis. The technology is well established. The hydration of the starch is facilitated by milling the grain, after which the finely divided starch becomes sufficiently hydrated when dispersed in water. Gelatinization involves hydration of the starch molecules, separation of the starch molecules, and rupture of some hydrogen bonds between glucose units. This process is governed by the type of starch, the time-temperature relationship, and the particle size. Temperatures, which generally range from about 60 to 100°C, may be lowered by grinding the starch to a finer particle.

Starch hydrolysis can be described by the following reaction:

$$(C_6H_{10}O_5)_n + nH_2O \longrightarrow n(C_6H_{12}O_6)$$

The fermentation process involves the conversion of simple hexose sugars to ethanol and carbon dioxide. The many possible hexose sugars differ by subtle structural alterations. The reaction is as follows:

$$C_6H_{12}O_6 \longrightarrow 2C_2H_5OH + CO_2$$

The conversion of molasses to ethanol does not require the starch hydrolysis steps necessary with grain fermentations. Therefore after molasses has been properly prepared by dilution, pH adjustment, nutrient addition, and pasteurization, it can be fermented directly. The fermentation process is quite similar to that discussed for the fermentation step in producing ethanol from corn. Of all the potential feedstocks that can be used to manufacture ethanol, sugar is the easiest to convert and is an important crop in developing and tropical countries. The main sources of sugar feedstocks for ethanol production are sugarcane, sugar beets, and sweet sorghum.

A major factor in the possible displacement of synthetic ethanol by fermentation ethanol for industrial application is the competitive cost-of-production economics of the synthetic versus fermentation process routes. Synthetic alcohol production by means of ethylene hydration involves the catalytic addition of water to ethylene. A common catalyst is an inert support impregnated with phosphoric acid. In general, pressures of about 1000 psi and temperatures of 300°C are used. The exothermic reaction is as follows:

$$CH_2{=}CH_2 + H_2O \longrightarrow C_2H_5OH$$

By-products, which are produced in small amounts, include ethers, aldehydes, higher hydrocarbons, higher alcohols, and ketones. The principal by-product is diethyl ether, which is generally recycled. The reactor operates at low conversions per pass, requiring a large recycle volume of unreacted ethylene. The reactor effluent is cooled and neutralized, and the condensed ethanol and water are separated from the unreacted gases, which are primarily unreacted ethylene. These gases are scrubbed with water. The condensed ethanol and water and scrubber liquid are combined and distilled to 95 percent purity by conventional distillation. To produce anhydrous ethanol, an ethanol-water azeotrope is fed to a dehydration distillation system. A block flow diagram is shown in Figure 2-5.

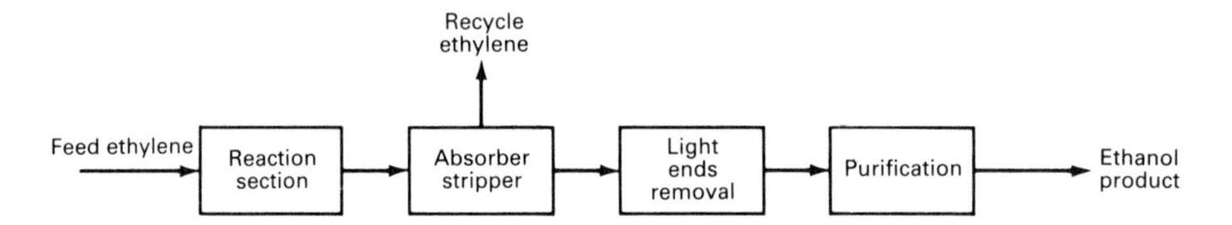

Figure 2.5 Production of ethanol from ethylene via direct hydration

Market

The future of ethanol use in fuel applications is dependent on trends in energy demand, supply, and pricing. Some of these trends have been discussed in the introduction. The demand for ethanol from the noncommunist countries was in excess of 9 million metric tons in 1983 and is generally projected to increase to over 16 million metric tons by 1990, mainly because of its increased use as fuel. South America is the largest ethanol-consuming region, followed by the United States. The chemical and solvent applications for ethanol are mature in most developed regions of the world, with only modest future growth expected. Regarding fuel uses, ethanol can be used as a fuel extender, neat fuel, and octane improver. In South America and the United States, government incentives for fuel ethanol production and consumption are offered to reduce dependence on imported oil. In South America the Proalcool program in Brazil is the most extensive fuel ethanol program in the world (3). The recent stability in oil availability and pricing has resulted in the reduction, postponement, or cancelation of fuel ethanol programs in many regions of the world.

World trade in ethanol is expected to at least double by the end of the decade. The major exporter will be Brazil, probably followed by Saudi Arabia. The price varies substantially in many areas because of tariffs and government influence to protect local industry.

Major producers of ethanol in the United States are shown in Table 2-15 (10). In addition, over 20 producers have capacities of, for the most part, less than 10 million gallons per year, primarily for fuel uses.

TABLE 2-15 Major U.S. Producers of Ethanol

Producer	Location	Estimated Capacity mm gal/yr	Estimated Capacity m MT/yr
Amber Labs	Juneau, Wisconsin	10	30
American Diversified	Hastings, Nebraska; Hamburg, Iowa	15	45
ADM	Cedar Rapids, Iowa; Decatur, Illinois; Peoria, Illinois	300	895
Eastman	Longview, Texas	25	75
Georgia Pacific	Bellingham, Washington	6	18
Grain Processing	Muscatine, Iowa	60	179
Kentucky Agriculture Energy	Franklin, Kentucky	21	63
Midwest Solvents	Atchison, Kansas	22	66
	Pekin, Illinois	9	27
Nabisco Brands	Clinton, Iowa	8	24
National Distillers	Tuscola, Illinois	66	197
New Energy	South Bend, Indiana	53	159
Pekin Energy	Pekin, Illinois	60	179
A. E. Staley	Loudon, Tennessee	60	179
Shepherd Oil	Jennings, Louisiana	35	105
Southport Ethanol	Southport, Ohio	60	179
Union Carbide	Texas City, Texas	120	358
	TOTAL	930	2778

Major Western European producers of synthetic ethanol are shown in Table 2-16 (11). Japan has five producers of synthetic alcohol. The two major producers are Japan Synthetic Alcohol, with a facility at Kawasaki, and Japan Ethanol, with a facility at Yokkaichi. Each plant has an estimated capacity of about 13 million gallons per year (40 m MT/yr).

TABLE 2-16 Major Western European Producers of Ethanol

Producer	Location		Estimated Capacity mm gal/yr	m MT/yr
Soc. Eth. Synth.	Lillebonne, France		42	125
Chem. Werke Huls	Wanne-Eickel, W. Germany		44	130
EC Erdolchemie	Koln, W. Germany		8	24
BP Chemicals	Grangemouth, U.K.		52	155
	Port Talbot, U.K.		52	155
		TOTAL	198	589

Economics

Typical production costs of ethanol by ethylene hydration are as follows:

Capacity: 440 mm lb/yr (67 mm gal/yr; 200 m MT/yr)
Capital costs: BLCC, $45 mm; OSBL, $18 mm; WC, $34 mm

	¢/lb	$/gal	$/MT	%
Raw materials[a]	13.9	0.91	306	61
Utilities	5.3	0.35	117	23
Operating costs	1.0	0.07	22	4
Overhead costs	2.8	0.18	62	12
Cost of production	23.0	1.51	507	100
Transfer price	27.3	1.79	602	

[a] Ethylene at 22¢/lb

ETHYLENE OXIDE AND ETHYLENE GLYCOL

The principal end uses for ethylene are ethylene oxide, ethylene glycol, and polyethylene. Well over half of the ethylene oxide is used in the production of ethylene glycol, and almost two-thirds of the ethylene glycol is used as antifreeze in motor vehicles and in the manufacture of polyester materials. Rising prices for ethylene are transmitted directly to the cost of the derivatives. A major change would occur if the glycol could be manufactured directly from the ethylene rather than going through the oxide.

Technology

Originally ethylene glycol was manufactured by the chlorohydrin process, involving the addition of hypochlorous acid to ethylene. The chlorohydrin intermediate was then hydrolyzed to ethylene glycol. The hypochlorous acid was readily obtained by the reaction of chlorine with aqueous calcium hydroxide.

In addition to the glycol, the chlorohydrin process also produced ethylene dichloride as a by-product. The consumption of chlorine ultimately resulted in the replacement of the process by the vapor-phase catalytic oxidation of ethylene using oxygen or air. Although direct oxidation yields are generally somewhat lower than yields via chlorohydrin, the overall economics favor direct oxidation, which is the process of choice in virtually all developed areas of the world.

The direct oxidation vapor-phase process operates at about 200 to 300°C and 150 to 450 psia pressure using a silver oxide catalyst. Generally, selectivities to ethylene oxide are close to 75 percent. As the costs of ethylene have risen, the selectivity of the direct oxidation process has become a major consideration in the economic viability of the process, and research and development continue toward further improvement of yields. In general, this improvement has been the result of increasingly selective catalyst systems, which minimize the oxidation of ethylene to carbon dioxide and water. Both the ethylene reaction rate and purge losses increase with higher concentrations of ethylene and oxygen. In different versions of the process, the ethylene concentration in the reactor inlet will vary from 5 to 40 volume percent, and inlet oxygen concentrations will generally be in the 5 to 9 percent range. Air-based reactors operate at the low end of the range, and oxygen-based reactors at the high end. The reactions are as follows:

$$CH_2H_4 + O_2 \xrightarrow{Ag_2O} \overset{\diagup O \diagdown}{CH_2{-}CH_2} + CO_2 + H_2O$$

$$\overset{\diagup O \diagdown}{CH_2{-}CH_2} \xrightarrow{H_2O} HOCH_2CH_2OH + HOCH_2CH_2OCH_2CH_2OH$$
$$+ HOCH_2CH_2OCH_2CH_2OCH_2CH_2OH$$

The ethylene glycol is generally obtained by noncatalytic thermal hydrolysis in the presence of an excess of water. A block flow diagram of a typical process is shown in Figure 2-6.

In addition to the ethylene glycol, di- and triethylene glycols are obtained as by-products. Conversion of the oxide to glycols is essentially complete, with selectivities of the ethylene glycol close to 95 percent. It is apparent from this

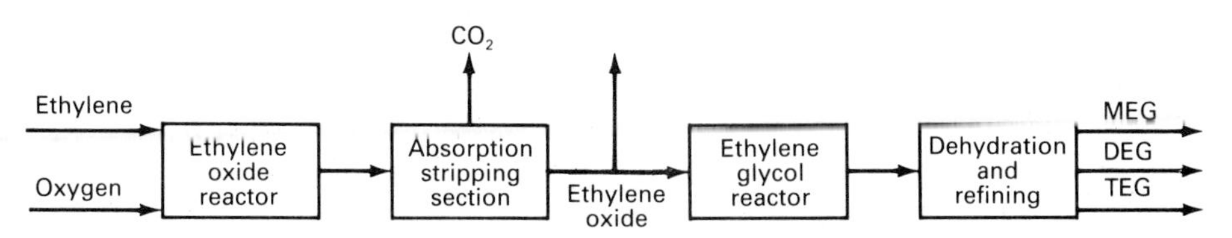

Figure 2.6 Production of ethylene oxide and ethylene glycol

route that selectivity to the ethylene oxide from the ethylene is the key to further improvement in the economics. Efforts have been long under way to develop a process for production of ethylene glycol directly, without going through the oxide or even starting with ethylene. Most of this work is going on in the United States, Western Europe, and Japan and includes the following routes:

1. Acetoxylation of ethylene: Production of mono- and diacetates followed by hydrolysis to ethylene glycol and acetic acid, which is recycled. Yields are relatively high, but severe corrosion problems exist. In 1978 Oxirane, now totally owned by Arco, started up an 800 million pound per year facility of this type in Channelview, Texas; but after severe corrosion problems resulted in discontinuous operation, the unit was shut down at the end of 1979.
2. Direct oxidation to ethylene glycol: A single-stage oxidation of ethylene to the glycol. A thallium catalyst system has been tried.
3. Ethylene glycol from synthesis gas: Involves an equimolar mixture of carbon monoxide and hydrogen to form the glycol. A process suggested by Union Carbide employs a complex rhodium carbonyl catalyst and operates at high pressures.
4. Ethylene glycol from formaldehyde, hydrogen, and carbon monoxide: In the initial step the three components are reacted with a hydrogen fluoride catalyst to form glycolic acid, which is then esterified with methanol. The final step involves the hydrogenation of the ester to form the glycol and by-product methanol, which is recycled. Many companies have been involved in similar technologies.
5. Ethylene glycol from ethylene by biotechnology: Several companies are working on a direct biotechnical route to the glycol using immobilized enzymes that ingest the ethylene and convert it to the glycol.

At present virtually all ethylene glycol is produced by the hydration of ethylene oxide, but many authorities feel that an economic route directly to the glycol has a good chance of being a commercial reality before the end of the next decade—probably from synthesis gas. Longer term, the biotechnological route is generally expected to become established.

Market

U.S. ethylene oxide production was 5.5 million pounds in 1983 and approximately the same level in 1984, with a projection in 1988 of about 6.3 million pounds (5). Most of this material was used in the production of ethylene glycol, with over 85 percent of the glycol going to the automobile antifreeze and polyester material market. The use of the glycol in polyester includes the production of fibers, film, and polyethylene terephthalate resins. One area of substantial expected growth is in the use of PET bottles (7). Ethylene glycol production in

TABLE 2-17 Uses of Ethylene Oxide and Ethylene Glycol

Application	Ethylene Oxide, % Used	Application	Ethylene Glycol % Used
Ethylene glycol	62	Polyester fiber and film	45
Surfactants	12	Antifreeze	39
Glycol ethers	7	PET bottles	5
Other, including exports	12	Other, including exports	11

the United States was about 4.4 billion pounds in 1983 and in 1984 and is projected to exceed 5 billion pounds by 1988. Some details of the uses of the oxide and glycol are shown in Table 2-17.

Worldwide capacity for ethylene oxide for 1983 was about 18 billion pounds, with Western and Eastern Europe and Japan following the United States, which has a total capacity of about 8 billion pounds. The next few years will show increases from the expected start-ups of several world-scale facilities in Saudi Arabia and Asian and African nations. Current producers in the United States are shown in Table 2-18 (5,6,8).

Major producers in Western Europe are shown in Table 2-19 (10). The four major ethylene glycol producers in Japan are shown in Table 2-20. The capacities shown include higher glycols.

TABLE 2-18 Current U.S. Producers of Ethylene Oxide and Ethylene Glycol

Producer	Location	Estimated Oxide Capacity		Estimated Glycol Capacity	
		mm lb/yr	m MT/yr	mm lb/yr	m MT/yr
BASF	Geismar, Louisiana	480	218	360	163
Celanese	Clear Lake, Texas	450	204	550	249
Dow Chemical[b]	Freeport, Texas	280	127	260	118
	Plaquemine, Louisiana	450	204	450	204
Eastman	Longview, Texas	190	86	190	86
ICI Americas	Bayport, Texas	500	227	500	227
Northern Petr.	Morris, Illinois	230	104	200	91
Olin	Brandenburg, Kentucky	110	50	40	18
PD Glycol[a]	Beaumont, Texas	455	206	620	281
Shell	Geismar, Louisiana	800	363	500	227
SunOlin	Claymont, Delaware	100	45		
Texaco	Port Neches, Texas	700	317	500	227
Union Carbide	Ponce, P.R.	630	286	650	295
	Seadrift, Texas	950	431	850	385
	Taft, Louisiana	1300	590	1400	635
	TOTAL	7625	3458	7070	3206

[a]PD Glycol is a joint venture of PPG and DuPont.
[b]Dow shut down its facilities at Freeport, Texas in 1984.

TABLE 2-19 Major Western European Producers of Ethylene Oxide and Ethylene Glycol

Producer	Location	Oxide Capacity		Glycol Capacity	
		mm lb/yr	m MT/yr	mm lb/yr	m MT/yr
BASF	Antwerp, Belgium	353	160	287	130
BP Chemicals	Zwijndrecht, Belgium	287	130	287	130
BP Chimie	Lavera, France	441	200	364	165
Ethylox	Gonfreville-L'Orcher, France	154	70	55	55
BASF	Ludwigshafen, W. Germany	331	150		
Chemische Werke Huels	Marl, W. Germany	331	150	198	90
EC Erdolchemie	Koln, W. Germany	331	150	243	110
Hoechst	Gendorf, W. Germany	265	120	243	110
	Kelsterbach, W. Germany	88	40	88	40
Anic	Gela, Italy	88	40	99	45
Montedison	Priolo, Italy	66	30	66	30
Riveda Sri	Brindisi, Italy	88	40	44	20
Dow Chemical	Terneuzen, Netherlands	276	125	265	120
Shell	Moerdijk, Netherlands	331	150	286	130
Ind. Quim. Asoc.	Tarragona, Spain	176	80	121	55
Berol Kemi	Stenungsund, Sweden	93	42	44	20
ICI	Wilton, U.K.	529	240	397	180
Shell	Carrington, U.K.	176	80	110	50
	TOTAL	4404	1997	3197	1450

TABLE 2-20 Major Japanese Producers of Ethylene Glycol

Producer	Location	Estimated Capacity mm lb/yr	m MT/yr
Japan Catalytic Chemical	Kawasaki	500	227
Mitsubishi Petrochemical	Kashima, Yokkaichi	410	186
Mitsui Petrochemical	Chiba, Sakai	450	204
Nisso Petrochemical	Goi, Yokkaichi	280	127
	TOTAL	1640	744

Economics

Typical production costs for ethylene oxide by direct oxidation and ethylene glycol by hydration of ethylene oxide are as follows:

Ethylene Oxide

Capacity: 400 mm lb/yr (181 m MT/yr)
Capital costs: BLCC, $51 mm; OSBL, $29 mm; WC, $34 mm

	¢/lb	$/MT	%
Raw materials[a]	20.2	445	79
Utilities	0.7	15	3
Operating costs	0.9	20	4
Overhead costs	3.7	82	14
Cost of production	25.5	562	100
Transfer price	31.5	695	

[a] Ethylene at 22¢/lb

Ethylene Glycol

Capacity: 325 mm lb/yr (147 m MT/yr)
Capital costs: BLCC, $16 mm; OSBL, $9 mm; WC, $22 mm

	¢/lb	$/MT	%
Raw materials[a]	25.6	564	77
Utilities	4.9	108	15
Operating costs	0.4	9	1
Overhead costs	2.2	49	7
Cost of production	33.1	730	100
By-product credit (higher glycols)	4.0	88	
Net cost of production	29.1	642	
Transfer price	31.5	695	

[a] Ethylene oxide at 31.5¢/lb

POLYETHYLENE

Low-density polyethylene was developed in the thirties and has grown and matured in both its market and technology. The polymer is produced in a multitude of grades and by many companies worldwide. The use of LDPE is concentrated in film and sheet applications, which account for most of the total production. In the United States most of the material is used in food packaging and garbage bags. Nonfood-packaging uses are even more important in Japan and Western Europe.

Growth in linear low-density polyethylene has mostly been at the expense of low-density polyethylene. For many applications the linear material has more desirable properties, and the data indicate a lower cost of production. The market for LLDPE is generally considered to be a replacement for LDPE and, to a lesser extent, for high-density polyethylene.

High-density polyethylene is basically a linear polymer with little side-chain branching and a specific gravity of 0.94 to 0.97, as compared with about 0.91 to 0.93 for LDPE. LDPE has a highly branched molecular structure and a relatively low crystallinity and melting point. LLDPE has a density between the high- and low-density polyethylene and fills the void in properties previously obtained by blending LDPE and HDPE.

Technology

Until fairly recently commercial production of LDPE was based on modifications of two high-pressure processes: autoclave and tubular polymerization. Operating pressures were approximately 1400 atmospheres. Several years ago new processes were developed that reduced pressures to about 20 atmospheres. The "Unipol" process, developed by Union Carbide, is a gas-phase, low-pressure process with a fluidized bed reactor. Resins produced are in the density range of 0.92 to 0.96, which overlaps the low- and high-density ranges. Dow Chemical used its own technology to convert an existing high-density polyethylene facility to LLDPE production in the late seventies. This technology is based on octene, whereas the Union Carbide process is based on butene. Many companies have developed their own low-pressure processes.

The early autoclave processes for LDPE production used oxygen as a reaction initiator. The use of unstable peroxides widened the range of properties, but yields per pass were relatively low—less than 20 percent. The autoclave process has back mixing of the hot reactants with cold incoming ethylene to maintain the stability of the reaction. This back mixing results in a range of residence times and therefore a range of molecular weights. Developments with baffled reactors and various initiators have permitted production of polymers with specific properties.

The tubular reactor process is usually preferred for the production of high-clarity film polymers because the reactor permits consistent molecular weight.

Conversion is somewhat higher than for the autoclave process. Peroxides are generally used as the initiator rather than oxygen.

The most significant development in polyethylene technology is the use of low-pressure technology. Many processes, using a variety of catalysts, have been developed, but the major process with the widest acceptance is the "Unipol" technology of Union Carbide. The primary advantage of the low-pressure over the high-pressure technology is economic. In general the low-pressure process offers significant savings in capital investment and utility costs. In addition, the process has the advantage of having wide flexibility, so much so that a facility can be designed that switches easily from low- to high-density polyethylene with a wide range of molecular weights and molecular weight distributions. If this technology is carried to its full potential, it may be possible to have a single facility that is flexible enough to produce all types of polyethylene. However, the resins produced by the low-pressure processes are, in general, materials with narrow molecular weight distributions and little branching—thus the name linear LDPE.

The high-pressure polymerization of ethylene is a free radical reaction and is initiated by a wide range of free radical agents, such as oxygen and organic peroxides. The process route can be considered to consist of free radical initiation, polymer chain preparation, and radical recombination. Side reactions can occur because of the highly reactive nature of the polyethylene free radical at the high temperatures involved. Chain termination eventually occurs, usually aided by the action of a chain moderator such as propane or butane.

The low-pressure process is a true catalyzed polymerization, unlike the high-pressure process. The Union Carbide process uses a supported catalyst containing chromium, titanium, and fluorine. The low-pressure polymerization of ethylene using a catalyst of reduced titanium chloride with aluminum alkyls results in HDPE, a highly crystalline polymer with a melting point of approximately 135°C. Pressures in the range of 200 to 500 psi are generally used.

The three basic processes that produce HDPE are the solution, slurry, and gas-phase technologies. The slurry and gas-phase technologies produce powders or flake, which is then formed into pellets; whereas the solution technology produces a pellet directly from the reactor. The product is slightly different with each process, and comonomers such as butene or hexene are commonly added. Commercial processes employ proprietary catalyst systems, which generally use catalysts in such small quantities that removal is rarely necessary. A typical block flowsheet for the production of polyethylene is shown in Figure 2-7.

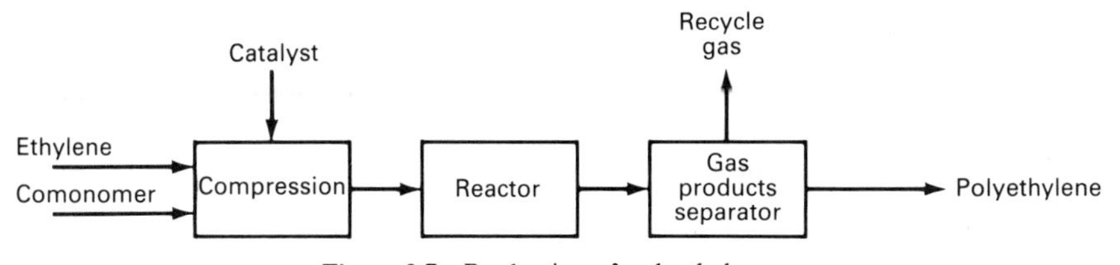

Figure 2.7 Production of polyethylene

Market

In the United States, Western Europe, and Japan, the major use for low-density polyethylene is for film and sheet applications. In the United States more than half of the film produced is used in food packaging and garbage bags. In the nonpackaging area are agricultural uses, such as mulch (ground cover); industrial liners, and construction uses, such as vapor barriers. Disposable trash bags are a large and growing market. Other than low price, which is very important, requirements are toughness, flexibility, and water impermeably. Another major area is injection molding applications, which are mostly accounted for by housewares and toys. Table 2-21 indicates U.S. production figures for all polyethylenes in 1983 and projected figures for 1992 in million pounds per year (3).

TABLE 2-21 U.S. Polyethylene Production for 1983 and 1992

Application	1983	1992
LDPE and LLDPE		
Film	4480	6667
Extrusion coating	560	645
Blow molding	40	60
Wire and cable	270	370
Injection molding	605	988
Other extruded products	80	100
All other products	815	1150
Domestic	6850	9980
Net export	900	(150)
TOTAL	7750	10,130
HDPE		
Blow molding	1775	3090
Injection molding	1140	2130
Pipe and conduit	470	860
Film	355	920
Sheet	75	156
Wire and cable	140	205
Other extruded products	40	41
All other products:	625	798
Domestic	4620	8200
Net export	1050	0
TOTAL	5670	8200
Total Polyethylene		
Total domestic	11,470	18,180
Net export	1950	(150)
TOTAL	13,420	18,030

TABLE 2-22 Major Installed and Planned Polyethylene Facilities

Producer	Country	Technology	Start-up	Capacity, m MT/yr		
				LDPE	LLDPE	HDPE
Esso	Canada	Union Carbide	1983		135	
Novacor	Canada	Union Carbide	1984		275	
Ipako	Argentina	Union Carbide	1981		120	
	Argentina	Arco	1982	75		
Polyolefinsa	Brazil	U.S.I.	1983	150		
Triunfo	Brazil	Ato	1984	100		
Singapore Poly.	Singapore	Sumitomo	1984	120		
Phillips	Singapore	Phillips	1984			80
Qatar/CdF	Qatar	CdF	1980	140		
Dow	Canada	Dow	1986		90	
Linoco	Libya	Carbide/ICI	1988	50	80	80
Exxon/Sabic	Saudi Arabia	Union Carbide	1985		260	
Mitsubishi/Sabic	Saudi Arabia	Union Carbide	1985		130	
Mobil/Sabic	Saudi Arabia	Union Carbide	1985		200	90
			TOTAL	635	1290	250

TABLE 2-23 Major U.S. Producers of LDPE and LLDPE

Producer	Location	Estimated Capacity	
		mm lb/yr	m MT/yr
Chemplex	Clinton, Iowa	415	188
Dow	Freeport, Texas	620	281
	Plaquemine, Louisiana	350	159
DuPont	Orange, Texas	465	211
	Victoria, Texas	240	109
Eastman Kodak	Longview, Texas	400	181
El Paso Products	Odessa, Texas	380	172
Exxon	Baton Rouge, Louisiana	580	263
Gulf	Cedar Bayou, Texas	620	281
	Orange, Texas	285	129
InterNorth	Morris, Illinois	560	254
Mobil	Beaumont, Texas	500	227
National Dist. & Chem.	Deer Park, Texas	560	254
	Tuscola, Illinois	165	75
	TOTAL	6140	2784

The influence of the new polyethylene capacity of the energy-rich nations will affect the decisions of U.S. producers for several years, and new capacity is not expected until later in the eighties. Major installed and planned facilities in plants designed primarily for polyethylene export are shown in Table 2-22.

Table 2-23 lists major producers of LDPE and LLDPE in the United States (1). Table 2-24 lists major producers of HDPE in the United States (1). Table 2-25 indicates major Western European producers of low-density polyethylene (2). Table 2-26 lists major Western European producers of high-density polyethylene (2).

TABLE 2-24 Major U.S. Producers of HDPE

Producer	Location	Estimated Capacity		LLDPE?
		mm lb/yr	m MT/yr	
Allied	Baton Rouge, Louisiana	700	317	
American Hoechst	Bayport, Texas	220	100	
Arco	Port Arthur, Texas	340	154	
Chemplex	Clinton, Iowa	290	132	
Dow Chemical	Freeport, Texas	150	68	Yes

TABLE 2-24 Major U.S. Producers of HDPE (*continued*)

Producer	Location	Estimated Capacity		LLDPE?
		mm lb/yr	m MT/yr	
	Plaquemine, Louisiana	220	100	Yes
DuPont	Bay City, Texas	440	200	
	Orange, Texas	205	93	
	Victoria, Texas	250	113	
Exxon	Mont Belvieu, Texas	600	272	Yes
Gulf	Orange, Texas	575	261	
Mobil	Beaumont, Texas	375	170	Yes
National Petrochem.	LaPorte, Texas	550	249	Yes
Phillips	Pasadena, Texas	1380	626	
Soltex Polymer	Deer Park, Texas	750	340	Yes
Amoco	Chocolate Bayou, Texas	350	159	
Union Carbide	Seadrift, Texas	800	363	Yes
	Taft, Louisiana	600	272	Yes
	TOTAL	8795	3989	

TABLE 2-25 Major Western European Producers of Low-Density Polyethylene

Producer	Location	Estimated Capacity		LLDPE?
		mm lb/yr	m MT/yr	
Danubia Olefinwerke	Schwechat, Austria	518	235	
BP Chemicals	Zwijndrecht, Belgium	243	110	
Essochem	Antwerp, Belgium	529	240	
	Zwijndrecht, Belgium	529	240	
Neste Oy	Kulloo, Finland	386	175	
Atochem	Balan, France	276	125	Yes
	Gonfreville-L'Orcher, France	364	165	
	Mont, France	176	80	
CdF Chimie	Carling, France	485	220	
	Dunkerque, France	221	100	Yes
COCHIME	Berre-L'Etang, France	221	100	
Copenor	Dunkerque, France	617	280	Yes
ICI	Fos-Sur-Mer, France	221	100	
BASF	Ludwigshafen, W. Germany	40	18	
EC Erdolchemie	Koln, W. Germany	662	300	
Rheinische Olefin.	Wesseling, W. Germany	882	400	
Ruhrchemie	Oberhausen, W. Germany	386	175	
Enichem Polimeri	Brindisi, Italy	397	180	
	Ferrara, Italy	243	110	

TABLE 2-25 Major Western European Producers of Low-Density Polyethylene (*continued*)

Producer	Location	Estimated Capacity		LLDPE?
		mm lb/yr	m MT/yr	
	Gela, Italy	397	180	
	Priolo, Italy	342	155	Yes
	Ragusa, Italy	265	120	
Dow Chemical	Terneuzen, Netherlands	461	209	Yes
DSM	Geleen, Netherlands	882	400	Yes
ICI	Rotterdam, Netherlands	165	75	
I/S Norpolefin	Saga, Norway	276	125	
SARL-EPSI	Sines, Portugal	265	120	
Alcudia	Puertollano, Spain	289	131	
	Tarragona, Spain	221	100	
Dow Chemical	Tarragona, Spain	276	125	Yes
Unifos Kemi	Stenungsund, Sweden	695	315	Yes
BXL Plastics	Grangemouth, U.K.	154	70	
	Wilton, U.K.	154	70	
Shell Chemicals	Carrington, U.K.	342	155	
	TOTAL	12,580	5703	

TABLE 2-26 Major Western European Producers of High-Density Polyethylene

Producer	Location	Estimated Capacity		LLDPE?
		mm lb/yr	m MT/yr	
Danubia Olefinwerke	Schwechat, Austria	300	136	
Dow Chemical	Tessenderlo, Belgium	185	84	
Petrochim	Antwerp, Belgium	496	225	
BP Chimie	Lavera, France	207	94	Yes
Soc. Ind. de Polyolef.	Gonfreville-L'Orcher, France	132	60	
Solvay & Cie	Sarralbe, France	309	140	
Hoechst	Frankfurt, W. Germany	309	140	
	Hurth-Knapsack, W. Germany	221	100	
	Munchsmunster, W. Germany	176	80	
Rheinische Olefin.	Wesseling, W. Germany	474	215	
Ruhrchemie	Oberhausen, W. Germany	309	140	
Vestolen	Gelsenkirchen, W. Germany	397	180	
Enichem Polimeri	Brindisi, Italy	386	175	

TABLE 2-26 Major Western European Producers of High-Density Polyethylene (*continued*)

Producer	Location	Estimated Capacity mm lb/yr	m MT/yr	LLDPE?
	Porto-Torres, Italy	154	70	
Solvay & Cie	Rosignano-Solvay, Italy	198	90	
Dow Chemical	Terneuzen, Netherlands	198	90	Yes
DSM	Geleen, Netherlands	331	150	Yes
I/S Norpolefin	Saga, Norway	132	60	
SARL-EPSI	Sines, Portugal	132	60	
CALATRAVA	Puertollano, Spain	143	65	
	Tarragona, Spain	132	60	
Dow Chemical	Tarragona, Spain	110	50	Yes
Tarragona Quimica	Reus, Spain	176	80	
Unifos Kemi	Stenungsund, Sweden	628	285	Yes
BP Chemicals	Grangemouth, U.K.	326	148	
	TOTAL	6561	2977	

Major producers in Japan of LDPE, LLDPE, and HDPE and their estimated capacities are shown in Tables 2-27, 2-28, and 2-29, respectively. Note that in many cases companies are converting their HDPE capacity to LLDPE capacity as the market warrants.

TABLE 2-27 Major Japanese Producers of LDPE

Producer	Location	Estimated Capacity mm lb/yr	m MT/yr
Asahi-Dow	Kawasaki, Mizushima	285	129
Mitsubishi Chemical	Mizushima	240	109
Mitsubishi Petrochemical	Kashima, Yokkaichi	750	340
Mitsui Petrochemical	Chiba, Ohtake	385	175
Nippon Petrochemical	Kawasaki	210	95
Nippon Unicar	Kawasaki	570	259
Showa Denko	Oita	275	125
Sumitomo	Chiba	440	200
Toyo Soda	Nanyo, Yokkaichi, Chiba	680	308
	TOTAL	3835	1740

TABLE 2-28 Major Japanese Producers of LLDPE

Producer	Location	Estimated Capacity mm lb/yr	m MT/yr
Idemitsu	Chiba	130	59
Mitsubishi Petrochemical	Kashima, Yokkaichi	175	79
Mitsui Petrochemical	Chiba, Ohtake	200	91
Nippon Unicar	Kawasaki	165	75
Showa Denko	Oita	130	59
	TOTAL	679	304

TABLE 2-29 Major Japanese Producers of HDPE

Producer	Location	Estimated Capacity mm lb/yr	m MT/yr
Asahi Chemical	Mizushima	240	109
Chisso	Chiba, Goi	100	45
Idemitsu	Chiba	165	75
Mitsubishi Chemical	Mizushima	140	63
Mitsubishi Petrochemical	Yokkaichi	75	34
Mitsui Petrochemical	Ohtake, Chiba	550	249
Nippon Polyethylene	Chiba	150	68
Shin Daikyowa	Yokkaichi	140	63
Showa Denko	Oita	330	150
Tonen	Kawasaki	90	41
	TOTAL	1980	897

Economics

Typical production costs for LDPE for a tubular reactor process and for HDPE and LLDPE via gas-phase technology are as follows:

LDPE

Capacity: 330 mm lb/yr (150 m MT/yr)
Capital cost: BLCC, $60 mm; OSBL, $21 mm; WC, $34 mm

	¢/lb	$/MT	%
Raw materials[a]	22.7	501	74
Utilities	2.5	55	8
Operating costs	1.0	22	3
Overhead costs	4.5	99	15
Cost of production	30.7	677	100
Transfer price	38.1	840	

[a] Ethylene at 22¢/lb

HDPE

Capacity: 220 mm lb/yr (100 m MT/yr)
Capital cost: BLCC, $47 mm; OSBL, $20 mm; WC, $24 mm

	¢/lb	$/MT	%
Raw materials[a]	23.6	520	71
Utilities	2.1	46	6
Operating costs	1.6	35	5
Overhead costs	6.1	135	18
Cost of production	33.4	736	100
Transfer price	43.4	957	

[a] Ethylene at 22¢/lb

LLDPE

Capacity: 440 mm lb/yr (200 m MT/yr)
Capital cost: BLCC, $62 mm; OSBL, $25 mm; WC, $43 mm

	¢/lb	$/MT	%
Raw materials[a]	23.9	527	80
Utilities	0.9	20	3
Operating costs	1.0	22	3
Overhead costs	4.0	88	14
Cost of production	29.8	657	100
Transfer price	35.7	787	

[a] Ethylene at 22¢/lb

ALPHAOLEFINS

Alphaolefins is a term generally used for a linear mono-olefin having the general chemical structure $RCH{=}CH_2$. Commercial production is usually accomplished by either the cracking of normal paraffin waxes or by the catalytic oligomerization of ethylene. Major end applications include plasticizer and detergent range alcohols via the oxo process, alphaolefin sulfonates, synthetic lubricants, and thermoplastic comonomers.

Technology

At present the major commercial routes to alphaolefins are wax cracking and polymerization of ethylene. Until the mid-sixties substantial production was from the dehydration of alcohols, but relatively low yields and increasing feedstock costs have essentially made this process obsolete. Wax cracking produces materials having both odd and even carbon chains, whereas ethylene polymerization, usually with a Ziegler-type catalyst, produces materials with only even chain lengths.

The wax cracking process generally uses a feedstock consisting primarily of C_{20} to C_{34} straight-chain paraffins. The paraffins are obtained from fractionation of waxy crude oil, which is generally followed by a deoiling process. As mentioned previously, the alphaolefins product obtained contains both odd and even carbon chain lengths. They are usually contaminated with such materials as branched olefins, diolefins, and paraffins.

The polymerization technology requires polymer-grade ethylene as the feedstock. The synthesis of alphaolefins from ethylene involves the addition of ethylene to an aluminum alkyl compound catalyst. Oligomerization then occurs. Chain growth is terminated when the alkyl group is displaced by an ethylene molecule. With triethylaluminum as the compound, the synthesis proceeds by reductive alkylation of aluminum as follows:

$$Al + \tfrac{3}{2}H_2 + 2Al(C_2H_5)_3 \longrightarrow 3Al(C_2H_5)_2H$$

$$3Al(C_2H_5)_2H + 3C_2H_4 \longrightarrow 3Al(C_2H_5)_3$$

The growth of the chain continues by the addition of ethylene to give a series of ligands, where R_1, R_2, and R_3 are oligomers of ethylene of either the same or varying chain length.

$$Al(C_2H_5)_3 + C_2H_4 \longrightarrow \begin{array}{c} R_1 \\ | \\ Al-R_2 \\ | \\ R_3 \end{array}$$

The ethylene is attached between the aluminum and the alkyl groups. The reaction is generally carried out at at least 100°C and 200 atmospheres. A block flow diagram for a typical process is shown in Figure 2-8.

Improvements in this process have focused on obtaining higher catalyst efficiency, higher selectivity to the desired chain length, and separation and recovery. The wax cracking process is still used in the United States by Chevron and in Europe by Shell. Ethylene polymerization is used in the United States by Ethyl, Gulf, and Shell and in Japan by Mitsubishi Chemical.

Market

At present four major producers of alphaolefins are in the United States. These producers and their locations and estimated capacities are shown in Table 2-30 (2). All but Chevron use the ethylene oligomerization process.

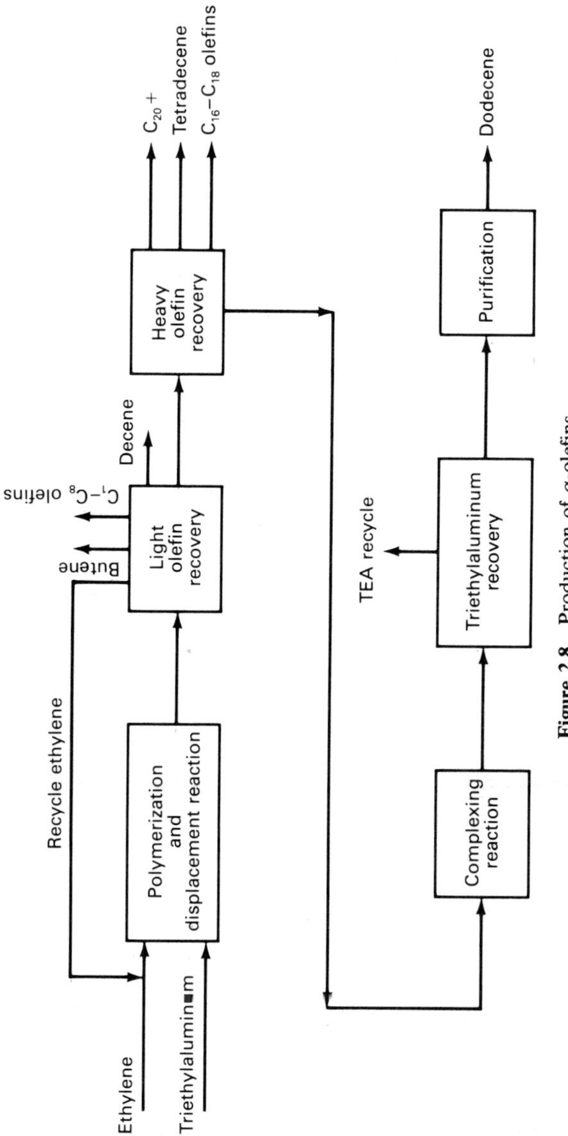

Figure 2.8 Production of α-olefins

TABLE 2-30 Major U.S. Producers of Alphaolefins

Producer	Location	Estimated Capacity	
		mm lb/yr	m MT/yr
Ethyl	Pasadena, Texas	> 800	> 363
Gulf	Cedar Bayou, Texas	200	91
Shell	Geismar, Louisiana	230	104
Chevron	Richmond, California	90	41
	TOTAL	> 1320	> 599

The small merchant market for alphaolefins in Western Europe ties its production with that of oxoalcohols. The only producer, Shell, uses wax cracking, but expansion using ethylene polymerization has been used at a facility at Stanlow, England, with a capacity of 375 million pounds per year (170,000 MT/yr) (3). Another facility is reportedly planned for the late eighties in the Netherlands (3). The wax cracking facilities are at Berre-L'Etang, France, with a capacity of about 200 million pounds per year (90,000 MT/yr) and at Pernis, Netherlands, with a capacity of about 350 million pounds per year (160,000 MT/yr).

In Japan Mitsubishi Chemical has a 55 million pound per year (25,000 MT/yr) facility at Mizushima based on ethylene, and Mitsubishi Petrochemical can produce approximately the same amount by the dehydrogenation of paraffins.

Alphaolefins have unusual pricing characteristics when compared with other major petrochemical intermediates. They are less influenced by cost trends for production, because they are used in several different market areas and are substantially influenced by the cost of producing the material from alternative feedstocks, the different routes for the production of the derivatives, and the prices for products competing with the alphaolefin derivatives. Butene-1 and octene-1 comonomers are used in the production of LLDPE at an average proportion of 9 pounds per 100 pounds of the plastic. Both can be produced by ethylene oligomerization. The production of LLDPE is one of the largest potential markets for alphaolefins. The emerging market for synthetic lubricants is another potentially strong market for the material. The use for alphaolefin sulfonates is generally considered to have the fastest growth rate predicted for major surfactants (4).

Economics

Typical production costs for alphaolefins are as follows:

Capacity: 300 mm lb/yr (136 m MT/yr)
Capital costs: BLCC, $37 million; OSBL, $18 million; WC, $31 million

	¢/lb	$/MT	%
Raw materials[a]	22.7	500	75
Utilities	3.4	75	11
Operating costs	0.8	18	3
Overhead costs	3.2	71	11
Cost of production	30.1	664	100
By-product credit (high MW olefins)	1.2	26	
Net cost of production	28.9	637	
Transfer price	34.4	759	

[a] Ethylene at 22¢/lb

REFERENCES

Acetaldehyde

1. *1984 Directory of Chemical Producers* — United States SRI International.
2. *1984 Directory of Chemical Producers — Western Europe*, SRI International.
3. "Petrochemical Handbook '83," *Hydrocarbon Processing*, November 1983.
4. *Kirk-Othmer Encyclopedia of Chemical Technology*, Third Edition, Vol. 1 New York: Wiley, pp. 97–112.
5. Chemical Profile, "*Chemical Marketing Reporter*, May 30, 1982.
6. *JCW Chemicals Guide 82/83*, Japan: The Chemical Daily Co. Ltd.

Ethylene Dichloride and Vinyl Chloride Monomer

1. List, H. L. "Vinyl Chloride Monomer—A Status Report," *International Petrochemical Developments*, Vol. 2, No. 14, July 15, 1981.
2. "Shell Assays Long-Term Goals; Closes Norco Ethylene and VCM," *Chemical Marketing Reporter*, April 25, 1983.
3. "Chemical Profiles," *Chemical Marketing Reporter*, June 20, 1983.
4. *1984 Directory of Chemical Producers—United States*, SRI International.
5. "Petrochemical Handbook '83," *Hydrocarbon Processing*, November 1983.
6. *1984 Directory of Chemical Producers—Western Europe*, SRI International.
7. *Kirk-Othmer Encyclopedia of Chemical Technology*, Third Edition, Vol. 23, New York: Wiley, 1984, pp. 865–85.
8. *JCW Chemicals Guide 82/83*. Japan: The Chemical Daily Co. Ltd., March 1982.
9. *Chemical Industry Yearbook*, Second Edition. Surrey, England: Industrial Press, 1984.

Polyvinyl Chloride

1. List, H. L. "Polyvinyl Chloride—A Status Report," *International Petrochemical Developments,* Vol. 2, No. 16, August 1, 1981.
2. "Chemical Profile," *Chemical Marketing Reporter,* June 27, 1983.
3. Brown, A. "PVC Competition Gets Fierce, Fiercer, Fiercest," *Chemical Business,* March 9, 1981, pp. 37–43.
4. "Polyvinyl Chloride Shakeout Gains Momentum," *Chemical and Engineering News,* April 5, 1982, p. 20.
5. "PVC—A Plastic in Turmoil," *Chemical Engineering,* May 17, 1982, pp. 49–51.
6. "Polyvinyl Chloride Looks Like a Winner Again," *Chemical Week,* May 4, 1983, pp. 26–27.
7. "Polyvinyl Chloride Catches Its Breath," *Chemical Week,* November 30, 1983, pp. 20–22.
8. "Look, the PVC Makers Are Smiling Again," *Chemical Week,* April 4, 1984, pp. 24–25.
9. Bayer, F. "PVC: Five More Years on the Tightrope," *Chemical Business,* January 1984, pp. 11–13.
10. "Petrochemical Handbook '83," *Hydrocarbon Processing,* November 1983.
11. *1984 Directory of Chemical Producers—Western Europe,* SRI International.
12. "The PVC Paradox: Higher Demand, Lower Prices," *Chemical Week,* October 24, 1984, pp. 10–12.
13. *1984 Directory of Chemical Producers—United States,* SRI International.
14. *Kirk-Othmer Encyclopedia of Chemical Technology,* Third Edition, Vol. 23. New York: Wiley, 1984, pp. 886–936.
15. *JCW Chemicals Guide 82/83.* Japan: The Chemical Daily Co. Ltd., March 1982.
16. "Materials '85," *Modern Plastics,* January 1985, pp. 61–71.
17. *Chemical Industry Yearbook,* Second Edition. Surrey, England: Industrial Press, 1984.

Vinyl Acetate

1. List, H. L. "Vinyl Acetate Update," *International Petrochemical Developments,* Vol. 4, No. 4, February 15, 1983.
2. Ehrler, J. L. and B. Juran, "VAM and Ac_2O by Carbonylation," *Hydrocarbon Processing,* February 1982, pp. 109–13.
3. "Chemical Profile," *Chemical Marketing Reporter,* May 23, 1983.
4. "Petrochemical Handbook '83," *Hydrocarbon Processing,* November 1983.
5. *1984 Directory of Chemical Producers—Western Europe,* SRI International.
6. "Key Chemicals—Vinyl Acetate," *Chemical and Engineering News,* September 24, 1984.
7. *1984 Directory of Chemical Producers—United States,* SRI International.
8. *Kirk-Othmer Encyclopedia of Chemical Technology,* Third Edition, Vol. 23. New York: Wiley, 1984, pp. 817–47.

9. *JCW Chemicals Guide 82/83.* Japan: The Chemical Daily Co. Ltd., March 1982.
10. *Chemical Industry Yearbook,* Second Edition. Surrey, England: Industrial Press, 1984.

Ethanol

1. List, H. L. "Ethanol Outlook," *International Petrochemical Developments,* Vol. 1, No. 1, November 15, 1980.
2. Ibid., "Fuel Ethanol—Update," *International Petrochemical Developments,* Vol. 3, No. 14, July 15, 1982.
3. Ibid., "Brazil's Proalcool Program," *International Petrochemical Developments,* Vol. 5, No. 4, February 15, 1984.
4. "Gasohol Makers Are Filling Up for the Long Haul," *Chemical Week,* June 27, 1984, pp. 8–9.
5. "Alcohol: The Fuel Source of the Future," *Brazil Trade and Industry,* November 1983, pp. 18–20.
6. "Brazil's Drive to Fuel Up with Alcohol," *Chemical Week,* May 11, 1983, pp. 33–34.
7. Weiss, L. H., and C. J. Mikulka. "Gasohol: A Realistic Assessment," *Chemical Engineering Progress,* June 1981, pp. 35–41.
8. Gale, G. "How Important is Gasohol?" *Chemical Business,* March 9, 1981, pp. 25–33.
9. Unzelman, G. H. "Problems Hinder Full Use of Oxygenates in Fuel," *Oil and Gas Journal,* July 2, 1984, pp. 59–65.
10. *1984 Directory of Chemical Producers—United States,* SRI International.
11. *1984 Directory of Chemical Producers—Western Europe,* SRI International.
12. *Kirk-Othmer Encyclopedia of Chemical Technology,* Third Edition, Vol. 9. New York: Wiley, 1984, pp. 338–80.
13. *JCW Chemicals Guide 82/83.* Japan: The Chemical Daily Co. Ltd., March 1982.
14. "Boom in Fuel Ethanol Begets New Capacity and Research," *Chemical Engineering,* January 21, 1985, pp. 38–41.
15. "Chemical Profile," *Chemical Marketing Reporter,* February 25, 1985.

Ethylene Oxide and Ethylene Glycol

1. Ozero, B. J. and J. V. Procelli. "Can Developments Keep Ethylene Oxide Viable," *Hydrocarbon Processing,* March 1984, pp. 55–61.
2. "Ethylene Glycol," *Chemical Week,* March 7, 1984, pp. 32–35.
3. "Two New Routes to Ethylene Glycol from Synthesis Gas," *Chemical and Engineering News,* April 11, 1983, pp. 41–42.
4. List, H. L. "Ethylene Oxide and Ethylene Glycol," *International Petrochemical Developments,* Vol. 2, No. 7, April 1, 1981.
5. "Chemical Profiles," *Chemical Marketing Reporter,* February 6, 1984.
6. "Chemical Profiles," *Chemical Marketing Reporter,* February 13, 1984.
7. List, H. L. "PET for Container Use Keeps Growing," *International Petrochemical Developments,* Vol. 1, No. 1, November 15, 1980.

8. *1984 Directory of Chemical Producers—United States,* SRI International.
9. "Petrochemical Handbook '83," *Hydrocarbon Processing,* November 1983.
10. *1984 Directory of Chemical Producers—Western Europe,* SRI International.
11. "Key Chemicals—Ethylene Oxide," *Chemical and Engineering News,* December 17, 1984, p. 15.
12. "EG Makers Won't Lose Balance," *Chemical Marketing Reporter,* April 30, 1984.
13. *Kirk-Othmer Encyclopedia of Chemical Technology,* Third Edition, Vol. 9. New York: Wiley, 1984, pp. 432–71.
14. *Kirk-Othmer Encyclopedia of Chemical Technology,* Third Edition, Vol. 11. New York: Wiley, 1984, 933–56.
15. *JCW Chemicals Guide 82/83.* Japan: The Chemical Daily Co. Ltd., March 1982.
16. *Chemical Industry Yearbook,* Second Edition. Surrey, England: Industrial Press, 1984.

Polyethylene

1. *1984 Directory of Chemical Producers—United States,* SRI International.
2. *1984 Directory of Chemical Producers—Western Europe,* SRI International.
3. "Polyethylene Demand Recovery Expected to Continue in 1984," *Hydrocarbon Processing,* March 1984, pp. 17–19.
4. List, H. L. "Low Density Polyethylene," *International Petrochemical Developments,* Vol. 1, No. 2, December 1, 1980.
5. List, H. L. "Linear Low Density Polyethylene," *International Petrochemical Developments,* Vol. 3, No. 13, July 1, 1982.
6. "Flood of Saudi PE?," *Modern Plastics,* June 1984, pp. 12–14.
7. "Saudi Polyethylene's Importance to Worldwide Industry Shakeout Seen Overblown by US Partner," *Chemical Marketing Reporter,* April 16, 1984.
8. "Linear Polyethylene Conversion Processes Win New Adherents," *European Chemical News Review Supplement,* Special Report, November 21, 1983, pp. 13–15.
9. "Polymer Producers Face Wide Choice of LLDPE Process Options," *European Chemical News,* June 27, 1983, pp. 11–13.
10. "Signs of Optimism Emerging in the Polyethylene Industry," *European Chemical News,* June 20, 1983, pp. 14–17.
11. Crimmin, S. M. "How Competitive Is Linear Low Density Polyethylene," *Hydrocarbon Processing,* December 1982, pp. 75–78.
12. "Chemical Profile," *Chemical Marketing Reporter,* December 20, 1982.
13. "Chemical Profile," *Chemical Marketing Reporter,* December 13, 1982.
14. Mizuma, H. "Development and Application of Linear Low Density Polyethylene," *Chemical Economy and Engineering Review,* Vol. 14, No. 10 (No. 162), October 1982, pp. 25–30.
15. *Kirk-Othmer Encyclopedia of Chemical Technology,* Third Edition, Vol. 16. New York: Wiley, 1984, pp. 385–452.
16. *JCW Chemicals Guide 82/83.* Japan: The Chemical Daily Co. Ltd., March 1982.
17. "Materials '85," *Modern Plastics,* January 1985, pp. 61–71.
18. *Chemical Industry Yearbook,* Second Edition. Surrey, England: Industrial Press, 1984.

Alphaolefins

1. List, H. L. "Alphaolefins—A Status Report," *International Petrochemical Developments,* Vol. 2, No. 13, July 1, 1981.
2. *1984 Directory of Chemical Producers—United States,* SRI International.
3. "Alphaolefins Grow at Manageable Rate," *Chemical Marketing Reporter,* July 18, 1983.
4. "Surfactants—A Mature Market with Potential," *Chemical and Engineering News,* January 1, 1982.
5. *1984 Directory of Chemical Producers—Western Europe,* SRI International.

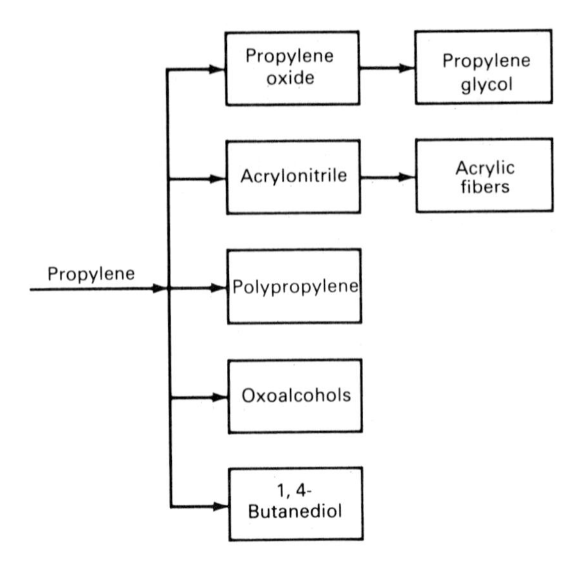

Propylene derivatives

CHAPTER THREE

Propylene Derivatives

PROPYLENE OXIDE AND PROPYLENE GLYCOL

Propylene oxide growth has essentially paralleled the growth of polyurethane resins and unsaturated polyesters. In the manufacture of the polyurethane resins, the intermediate is polypropylene glycol; whereas for the polyesters, the intermediate is propylene glycol. Until the late sixties virtually all propylene oxide was manufactured by the classical chlorohydrin route. In the late sixties a hydroperoxide-based process was developed jointly by Halcon International and Arco Chemical. The joint venture company was called Oxirane Corporation and is currently an Arco subsidiary.

Technology

Chlorohydrin route. The production of propylene oxide via chlorohydrin technology involves the addition of hypochlorous acid to propylene to form a chlorohydrin intermediate, which is then hydrolyzed to the epoxide according to the following reactions:

$$Cl_2 + H_2O \longrightarrow HOCl + HCl$$

$$CH_3CH{=}CH_2 + HOCl \longrightarrow CH_3\underset{\displaystyle OH}{\underset{|}{C}}HCH_2Cl + CH_3\underset{\displaystyle Cl}{\underset{|}{C}}HCH_2OH$$

$$CH_3\underset{\displaystyle OH}{\underset{|}{C}}HCH_2Cl + CH_3\underset{\displaystyle Cl}{\underset{|}{C}}HCH_2OH \xrightarrow{\text{base}} CH_3\underbrace{CHCH_2}_{O} + \text{Salt}$$

Generally, the base used is calcium hydroxide. This substance results in a dilute calcium chloride effluent, which presents environmental problems. For each pound of propylene oxide produced, about 40 pounds of aqueous effluent must be disposed of. This problem has led to several variations in the basic process. Sodium hydroxide could be used as the base and the sodium chloride brine recycled to a chlorine plant. A block flow diagram of a process with an integral caustic-chlorine plant is shown in Figure 3-1. However, this process will usually involve an evaporation step to maintain the required chlorine cell feed concentration in a closed cycle plant. If the chlorine plant is large, the evaporation step might be eliminated by adding solid salt. Other proposed modifications follow:

1. Kellogg-Bayer process: an electrochemical route that produces chlorohydrin at the anode. The chlorohydrin then passes through a porous diaphragm to the cathode, where propylene oxide and hydrogen are evolved.
2. University of Dortmund process: similar to the Kellogg-Bayer process except that a cation exchange membrane cell is used to separate the anolyte and catholyte.
3. Lummus process: uses a closed cycle of tertiary butyl hypochlorite in place of hypochlorous acid for the production of the chlorohydrin. Although treating the aqueous effluent from this process is costly, the modifications just described do not appear to have a major effect on the process economics.

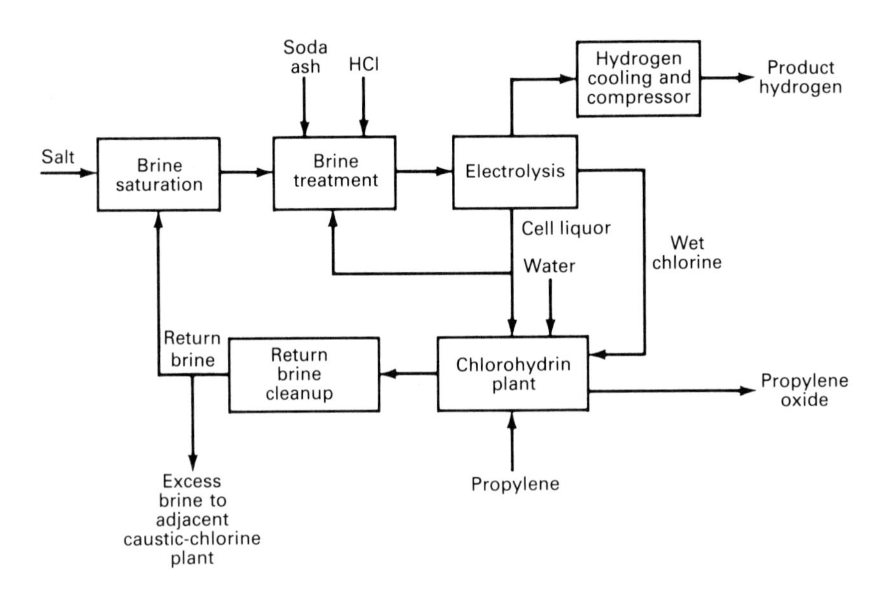

Figure 3.1 Production of propylene oxide—chlorohydrin process

Hydroperoxidation route. The Oxirane process involves the hydroperoxidation of isobutane or ethylbenzene. This step is followed by reaction with propylene to form the propylene oxide. For isobutane feed the reactions are as follows:

$$CH_3CH(CH_3)CH_3 + O_2 \longrightarrow CH_3C(CH_3)_2OOH$$

$$CH_3C(CH_3)_2OOH + CH_3CH{=}CH_2 \longrightarrow CH_3\overset{O}{CHCH_2} + CH_3C(CH_3)_2OH$$

The tertiary butyl alcohol can be sold or dehydrated to isobutylene. Isobutylene is a very desirable product used for the rapidly expanding production of methyl tertiary butyl ether, a gasoline octane improver. The reaction is as follows:

$$CH_3C(CH_3)_2OH \longrightarrow CH_3C(CH_3){=}CH_2 + H_2O$$

The comparable reactions starting with ethylbenzene are as follows:

$$C_6H_5CH_2CH_3 + O_2 \longrightarrow C_6H_5CH(CH_3)OOH$$

$$C_6H_5CH(CH_3)OOH + CH_3CH{=}CH_2 \longrightarrow CH_3\overset{O}{CHCH_2} + C_6H_5CH(CH_3)OH$$

The comparable dehydration of the alcohol will yield by-product styrene:

$$C_6H_5CH(CH_3)OH \longrightarrow C_6H_5CH{=}CH_2 + H_2O$$

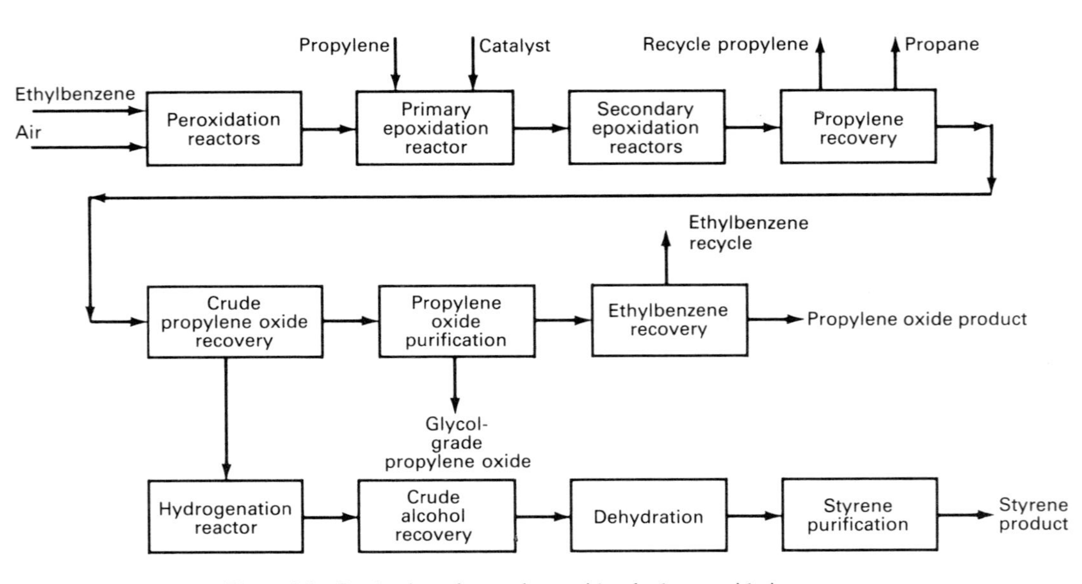

Figure 3.2 Production of propylene oxide—hydroperoxidation process

Obviously in these cases involving by-product production, the overall economics are highly dependent on the market and price for the by-product. A block flow diagram for the hydroperoxidation process starting with ethylbenzene is shown in Figure 3-2.

The technology for production of propylene glycol from propylene oxide is comparable with that for production of ethylene glycol from ethylene oxide and is discussed in Chapter 2.

Market

In the United States propylene oxide is currently produced primarily by Dow Chemical and Arco. Dow, whose production is mostly for captive use, employs the chlorohydrin technology. Arco uses the hydroperoxidation technology and primarily serves the merchant market. Both companies are considering expansions to cover the lost output from companies that have discontinued production. All of these companies used the chlorohydrin technology and include Union Carbide, Olin, and BASF-Wyandotte. Details of the U.S. production are shown in Table 3-1 (2,3).

U.S. demand for propylene oxide in 1983 was just under 2 billion pounds, and a demand of 2.4 billion pounds is projected for 1988. The key propylene oxide market for flexible polyurethane foams for furniture and transportation seat cushions is a mature market and is sensitive to swings in the economic cycles. In spite of the acceptance of RIM body parts by auto manufacturers, rigid urethanes have thus far not lived up to earlier projections of growth because of competitive alternatives such as styrene and phenolics. End uses and their percentages of the U.S. market are shown in Table 3-2 (3).

Demand for propylene glycol in 1984 was in excess of 460 million pounds in the United States and is projected to be over 530 million pounds by 1988. End uses and their percentages of the U.S. market are shown in Table 3-3 (8).

Major producers of propylene glycol in the United States are shown in Table 3-4 (2). Major producers of propylene oxide in Western Europe are shown in Table 3-5 (5). Most of these same companies are also producers of propylene

TABLE 3-1 U.S. Production of Propylene Oxide

		Estimated Capacity	
Producer	Location	mm lb/yr	m MT/yr
Arco	Bayport, Texas[a]	1020	463
	Channelview, Texas[b]	400	181
Dow	Freeport, Texas	950	431
	Plaquemine, Louisiana	440	200
	TOTAL	2810	1275

[a] Uses isobutane feedstock

[b] Uses ethylbenzene feedstock

TABLE 3-2 Propylene Oxide End Uses and Their Percentages of the U.S. Market

End Uses	Market %
Urethane polyether polyols	59
Propylene glycol	21
Dipropylene glycol	2
Other	9
Exports	9

TABLE 3-3 Propylene Glycol End Uses and Their Percentages of the U.S. Market

End Uses	Market %
Unsaturated polyester resins	44
Pet food	11
Pharmaceuticals and food	9
Tobacco humectant	8
Plasticizers	6
Cellophane	4
Other	10
Exports	8

TABLE 3-4 Major U.S. Producers of Propylene Glycol

Producer	Location	Estimated Capacity	
		mm lb/yr	m MT/yr
Arco	Bayport, Texas	250	113
Dow	Freeport, Texas	250	113
	Plaquemine, Louisiana	150	68
Olin	Brandenburg, Kentucky	70	32
Texaco	Port Neches, Texas	50	23
Union Carbide	South Charleston, W. Virginia	80	36
	TOTAL	850	385

TABLE 3-5 Major Western European Producers of Propylene Oxide

Producer	Location	Estimated Capacity	
		mm lb/yr	m MT/yr
ATOCHEM	Lavera, France	154	70
BASF	Ludwigshafen, W. Germany	143	65
Chemische Werke Huels	Marl, W. Germany	60	27
Dow Chemical	Stade, W. Germany	827	375
EC Erdolchemie	Koln, W. Germany	298	135
Montepolimeri	Priolo, Italy	88	40
Arco Chemie	Rotterdam, Netherlands	441	200
Shell Chemie	Moerdijk, Netherlands	276	125
Montoro-Empresa	Puertollano, Spain	99	45
	TOTAL	2386	1082

TABLE 3-6 Major Western European Producers of Propylene Glycol

Producer	Location	Estimated Capacity mm lb/yr	m MT/yr
ATOCHEM	Lavera, France	84	38
BASF	Ludwigshafen, W. Germany	88	40
Chemische Werke Huels	Marl, W. Germany	46	21
Dow Chemical	Stade, W. Germany	198	90
EC Erdolchemie	Koln, W. Germany	33	15
Montepolimeri	Priolo, Italy	26	12
Arco Chemie	Rotterdam, Netherlands	132	60
Alcudia	Puertollano, Spain	26	12
	TOTAL	633	288

glycol (including mono, di, and tri), as shown in Table 3-6 (5). Major producers of propylene oxide and glycol in Japan include Nisso Petrochemical, Asahi Denka, Asahi Glass, Mitsui Toatsu Chemicals, and Showa Denko (9). Note that propylene glycol production facilities can be used for ethylene glycol production, and therefore capacities are quite flexible.

Economics

Typical production costs are presented in the following tables for the production of propylene oxide via the chlorohydrin process, propylene oxide via the hydroperoxidation of isobutane, and propylene glycol by the hydration of propylene oxide.

Propylene Oxide Via Chlorohydrin

Capacity: 400 mm lb/yr (181 m MT/yr)
Capital cost: BLCC, $140 mm; OSBL, $90 mm; WC, $58 mm

	¢/lb	$/MT	%
Raw materials[a]	18.6	410	43
Utilities	14.0	309	32
Operating costs	2.1	46	5
Overhead costs	9.0	198	20
Cost of production	43.7	963	100
By-products[b]	(4.0)	(88)	
Net cost of production	39.7	875	
Transfer price	56.9	1255	

[a] Propylene at 20¢/lb

[b] Hydrogen, dichloropropylene, bischloroether

Propylene Oxide Via Hydroperoxidation

Capacity: 400 mm lb/yr (181 m MT/yr)
Capital cost: BLCC, $155 mm; OSBL, $93 mm; WC, $41 mm

	¢/lb	$/MT	%
Raw materials[a]	56.8	1252	74
Utilities	9.8	216	13
Operating costs	2.0	44	2
Overhead costs	8.2	181	11
Cost of production	76.8	1693	100
By-products[b]	(48.9)	(1078)	
Net cost of production	27.9	615	
Transfer price	46.5	1025	

[a] Propylene at 20¢/lb; isobutane at 15¢/lb
[b] Tertiary butanol, acetone, propane, butane

Propylene Glycol Via Hydration of Propylene Oxide

Capacity: 150 mm lb/yr (68 m MT/yr)
Capital cost: BLCC, $12.6 mm; OSBL, $5.1 mm; WC, $16.0 mm

	¢/lb	$/MT	%
Raw materials[a]	37.6	829	80
Utilities	5.7	126	12
Operating costs	0.8	18	2
Overhead costs	3.1	68	6
Cost of production	47.2	1041	100
By-products[b]	(1.0)	(22)	
Net cost of production	46.2	1019	
Transfer price	58.0	1279	

[a] Propylene oxide at 47¢/lb
[b] Dipropylene glycol

ACRYLONITRILE

Virtually all acrylonitrile is produced by propylene ammoxidation. The major licensor in the United States is Sohio Chemicals.

Technology

The propylene ammoxidation process employs a catalyst, and the reaction takes place at about 450°C using air as the oxidizing agent. By-product HCN and acetonitrile are also obtained. The reaction can be summarized as follows:

$$CH_2{=}CHCH_3 + O_2 + NH_3 \longrightarrow CH_2{=}CHCN + HCN + CH_3CN + H_2O + CO_2$$

The original catalyst used was bismuth phosphomolybdate supported on silica, but this catalyst was subsequently replaced by a more efficient antimony-uranium oxide catalyst. At present a new installation would probably employ a catalyst containing no uranium.

The process involves the introduction of propylene, ammonia, and air into a fluid bed reactor at about 2 atmospheres pressure. Conversion is virtually complete, and selectivity to acrylonitrile based on propylene is approximately 67 percent. Major reaction by-products are hydrogen cyanide and acetonitrile. The remainder of the process involves the separation of the components of the reactor effluent. A block flow diagram of the process is shown in Figure 3-3.

This process replaced the earlier technology involving reaction between acetylene and HCN primarily because of lower raw material costs. Raw material costs account for about three-fourths of the cost of production, but the value realized by the by-product HCN and acetonitrile can in some cases offset this expense. HCN is generally a valuable by-product, but acetonitrile does not normally enjoy a large demand and in some situations is even a disposal problem. Some patent disclosures have revealed other potential routes, including the reaction of methane and acetonitrile or the use of propane as an alternative feedstock, but the economic incentive at this time is insufficient to pursue commercialization.

Market

The major end application for acrylonitrile is for the manufacture of acrylic and modacrylic fibers. Acrylics are the synthetic fibers that have properties closest to that of wool. The monomer is also used for acrylonitrile-butadiene-styrene (ABS) resins. The market percentage of each end use is shown in Table 3-7 (2).

U.S. demand for acrylonitrile was about 2.1 billion pounds in 1983 and is projected to be about 2.5 billion pounds by 1988. The growth is expected to be in the use of ABS resins and for exports. In Western Europe, use of acrylonitrile for fibers accounts for close to 80 percent of total consumption. Production in Western Europe is comparable with that of the United States, and Japan's production is about half of that amount. Table 3-8 shows the major U.S. producers (2,3).

Major producers in Western Europe are shown in Table 3-9 (5). Most producers in Western Europe use Sohio technology, except for Norsolor of France and Border of the United Kingdom, which use PCUK technology, and Montedison of Italy, which uses its own technology.

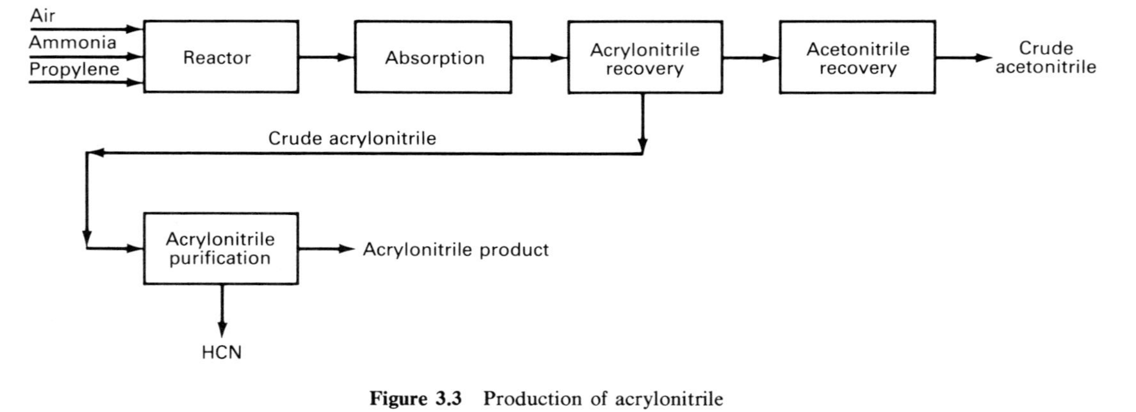

Figure 3.3 Production of acrylonitrile

TABLE 3-7 Percentages of End Uses of Acrylonitrile

Product	%
Acrylic and modacrylic fibers	32
ABS and SAN resins	15
Adiponitrile	9
Acrylamide	4
Miscellaneous	5
Exports	35

TABLE 3-8 Major U.S. Producers of Acrylonitrile

		Estimated Capacity	
Producer	Location	mm lb/yr	m MT/yr
Cyanamid	Fortier, Louisiana	265	120
DuPont	Beaumont, Texas	350	159
Monsanto	Chocolate Bayou, Texas	460	209
	Texas City, Texas	450	204
Sohio	Green Lake, Texas	400	181
	Lima, Ohio	350	159
	TOTAL	2275	1032

TABLE 3-9 Major Western European Producers of Acrylonitrile

		Estimated Capacity	
Producer	Location	mm lb/yr	m MT/yr
Chemie Linz	Enns, Austria	165	75
Norsolor	Saint-Avold, France	198	90
EC Erdolchemie	Koln, W. Germany	617	280
Hoechst	Munchsmunsher, W. Germany	198	90
ANIC	Assemini, Italy	154	70
	Gela, Italy	176	80
Montedison	Priolo, Italy	176	80
DSM	Geleen, Netherlands	331	150
Paular	Morell, Spain	165	75
Border	Grangemouth, U.K.	187	85
Monsanto	Middlesbrough, U.K.	441	200
	TOTAL	2808	1275

Major Japanese producers include Asahi Chemical Industries, Mitsubishi Chemical Industries, Mitsui Toatsu Chemicals, Nitto Chemical Industries, Showa Denko, and Sumitomo Chemical (7).

Because so much propylene in the United States comes from refinery sources, U.S. producers have a feedstock advantage over European and Japanese acrylonitrile manufacturers, who rely on coproduct propylene from ethylene manufacture. This advantage, in addition to the rationalization of acrylonitrile production overseas, contributes to the strong U.S. export business.

Toxicity

One of the most critical factors that will affect future demand for acrylonitrile is the problem of its carcinogenicity. The U.S. Food and Drug Administration withdrew its interim approval of nitrile food containers in the late seventies. This development spurred the use of the rival polyethylene terephthalate (PET) for this market, and acrylonitrile is unlikely to recover its market position. Many studies on the carcinogenicity of acrylonitrile have been made, but at this time its future status for use in food containers is uncertain.

Economics

Typical production costs for acrylonitrile are as follows:

Capacity: 660 mm lb/yr (272 m MT/yr)
Capital cost: BLCC, $117 mm; OSBL, $57 mm; WC, $67 mm

	¢/lb	$/MT	%
Raw materials[a]	27.6	609	83
Utilities	0.7	15	2
Operating costs	0.8	18	2
Overhead costs	4.3	95	13
Cost of production	33.4	737	100
By-product credit[b]	(3.0)	(66)	
Net cost of production	30.4	671	
Transfer price	39.1	862	

[a]Propylene at 20¢/lb
[b]HCN at 40¢/lb

ACRYLIC FIBERS

Acrylic fiber is produced by solution spinning of homopolymers or copolymers of acrylonitrile. Methyl acrylate and vinyl acetate are typical copolymers and are reported to decrease fiber shrinkage, thereby allowing the production of high-bulk material. Typically, the amount of comonomer used is 5 to 10 percent.

Fibers are generally classified as either acrylic fibers or modacrylic fibers, depending on their content of acrylonitrile.

Technology

The production of acrylonitrile was described in the preceding section. A block flow diagram for the production of acrylic fiber is shown in Figure 3-4.

Acrylic fibers are generally defined as long-chain synthetic polymers consisting of at least 85 percent by weight, of acrylonitrile units. DuPont in 1950 began producing acrylic continuous filament yarn and in 1952 began the production of acrylic staple. Yarn production was discontinued in 1956. Polyacrylonitrile resins can be produced by bulk polymerization, suspension polymerization, emulsion polymerization, and solution polymerization, but the major processes are suspension and solution polymerization. Suspension polymerization is accomplished by suspending small drops of acrylonitrile and comonomers in water using agitation and a stabilizer to prevent coalescense of the drops of monomer. A catalyst, soluble in the monomer, promotes the polymerization. The drops are converted into insoluble beads and are separated, washed, dried, milled, and redissolved in the spinning solvent.

In solution polymerization the acrylonitrile and comonomers are dissolved in an organic solvent or a concentrated aqueous solution of zinc chloride or other salts. An initiator triggers the reaction, and the polyacrylonitrile formed is soluble in the solvent. This process has the advantage of producing a polymer solution that can be directly used for spinning the fiber, which results in lower production costs. The comonomers used generally improve the fiber stretching and crimping properties and help in the dyeing of the product.

Acrylic fibers are either wet spun or dry spun. In wet spinning a polymer solution of about 20 percent polymer is extruded through a spinneret into a spinning bath. The bath contains a spinning solvent that has been diluted with water so that the fibers coagulate at a predetermined rate. In dry spinning the preheated solution is extruded through a spinneret into a column of circulating air, which is hotter than the boiling point of the solvent. The solvent evaporates, thereby causing the filaments to solidify. Wet spinning has the advantage that the remaining processing steps can be continuous. With both methods the fibers are treated to remove the residual solvent. Stretching and crimping may follow, and the fibers are thermally stabilized to set the crystalline structure.

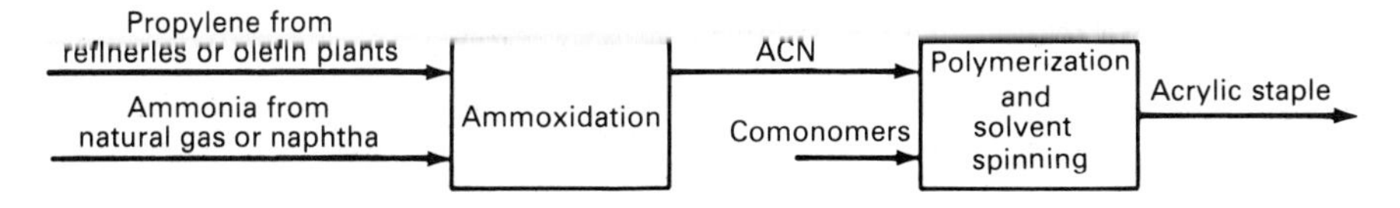

Figure 3.4 Production of acrylic staple

TABLE 3-10 Major U.S. Producers of Acrylic Fibers

Producer	Location	Estimated Capacity	
		mm lb/yr	m MT/yr
Am. Cyanamid	Milton, Florida	140	63
Badische	Williamsburg, Virginia	85	39
DuPont	Camden, S. Carolina	115	52
	Waynesboro, Virginia	165	75
Eastman	Kingsport, Tennessee	24	11
Monsanto	Decatur, Alabama	300	136
	TOTAL	829	376

Market

The success of acrylic fibers is largely due to its ability to be engineered to resemble most kinds of wool. Certain types can be woven into fabrics having the same feel and appearance as certain types of wool. The fibers dye easily in bright colors and have excellent resistance to shrinkage and weathering. The fibers have a lower tensile strength than polyester or nylon. Acrylic fiber consumption is centered in the apparel and home furnishings market.

Table 3-10 lists the major U.S. producers of acrylic fibers (1). Table 3-11 shows the major producers of acrylics in Western Europe (2).

TABLE 3-11 Major Western European Producers of Acrylic Fibers

Producer	Location	Estimated Capacity	
		mm lb/yr	m MT/yr
Courtaulds	Coquelles, France	132	60
Bayer	Dormagen, W. Germany	309	140
	Lingen, W. Germany	79	36
Hoechst	Kelheim, W. Germany	172	78
Vomvicryl	Lamia, Greece	22	10
Asahi	Ballina, Ireland	40	18
Anicfibre	Ottana, Italy	132	60
	Pisticci, Italy	110	50
	Porto-Torres, Italy	77	35
Snia Fibre	Cesano-Maderno, Italy	146	66
SIPA	Porto-Marghera, Italy	176	80
Fib. Sint. dePortugal	Lavradio, Portugal	51	23
Cyanenka	Prat-de-Llobregat, Spain	121	55
Montefibre	Miranda-de-Ebro, Spain	110	50
Courtaulds	Grimsby and Coventry, U.K.	209	95
Montefibre	Coleraine, U.K.	99	45
	TOTAL	1985	901

Economics

Typical production costs for acrylic staple from acrylonitrile via wet spinning are as follows:

Capacity: 100 mm lb/yr (45 m MT/yr)
Capital cost: BLCC, $110 mm; OSBL, $44 mm; WC, $30 mm

	¢/lb	$/MT	%
Raw materials[a]	53.3	1175	60
Utilities	6.0	132	7
Operating costs	8.2	181	9
Overhead costs	21.5	474	24
Cost of production	89.0	1962	100
Transfer price	135.2	2981	

[a] Acrylonitrile at 45¢/lb

POLYPROPYLENE

A major advantage of polypropylene is frequently its low price relative to other resins. This versatile polymer has favorable properties that have made it the third major polyolefin in the United States after low-density and high-density polyethylenes.

Technology

Before the development of the Ziegler-Natta catalysts, polymers of propylene were of limited commercial importance, since the products were amorphous and mostly low-molecular-weight material. The development of the new catalysts was the result of the work of Karl Ziegler of Germany and Giulio Natta of Italy, for which they shared a Nobel prize in chemistry. Use of the Ziegler-Natta catalysts permitted the production of crystalline high-molecular-weight polymer. There was substantial disagreement about patent ownership among the early developers of the process, and in early 1980 the U.S. courts ruled that Phillips Petroleum had priority of invention. Appeals were filed by DuPont, Montedison, and Standard Oil of Indiana. Montedison had previously, in 1954, been awarded a priority-of-invention patent for this catalyst.

All current commercial polypropylene processes use a Ziegler-Natta catalyst, which normally contains three components. The active ingredient is titanium trichloride. Together with an aluminum alkyl as cocatalyst and a polar compound, these components constitute the catalyst system. The polar compound, often an amine, minimizes the formation of undesirable noncrystalline propylene polymer. Most commercial processes for production of polypropylene involve one of four techniques for polymerization: slurry, solution, bulk, or vapor phase.

In slurry polymerization the propylene, in an inert hydrocarbon diluent at relatively low pressure and temperature, is contacted with the catalyst. Any amorphous polymer that forms dissolves in the diluent, making the reaction medium viscous and causing processing problems, mainly in heat exchange and mixing. Patented improvements in slurry polymerization have been developed by companies such as Shell and Montedison. A block flow diagram for a typical slurry polymerization process is shown in Figure 3-5.

In solution polymerization a sufficiently high temperature and pressure are maintained so that the polymer product is dissolved in the diluent. In this way, the increase in viscosity of the reaction medium can be controlled to some degree. Generally a lower concentration of catalyst is possible in solution polymerization, and for certain products the catalyst residue can be left in the polymer.

In bulk polymerization no inert diluent is used. Phillips Petroleum and El Paso Gas have patented variations of the bulk polymerization technique.

The vapor-phase polymerization process is used by BASF in West Germany. The polymer has better low-temperature impact strength than does conventional polypropylene and has the advantage of a catalyst content low enough to eliminate the need for removal of catalyst residues from the polymer. Many companies are developing improvements in the process, including Montedison and Hoechst. Shell and other companies have studied the use of fluidized bed reactors using gaseous propylene.

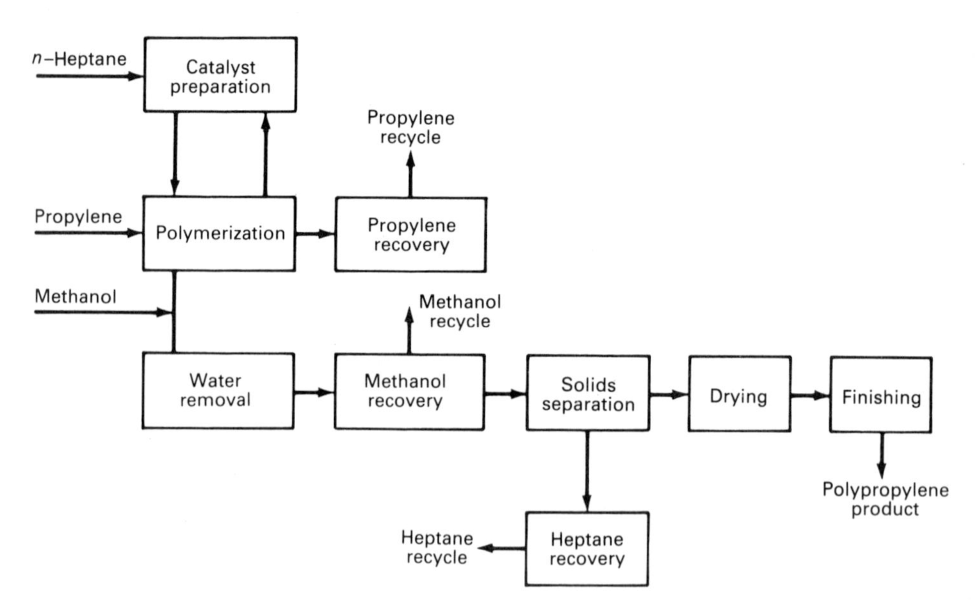

Figure 3.5 Production of polypropylene—slurry process

For the most part polypropylene prepared by slurry, solution, or bulk polymerization must be treated to remove catalyst residues and amorphous polymer. The catalyst residues discolor the polymer and lower its resistance to light and oxygen degradation. The presence of amorphous polymer lowers the tensile strength and adversely affects extrusion properties.

Technical improvements continue to be announced. Much of the work is in the development of high-selectivity, high-polymer-yield catalyst systems. A major new process has been announced by Union Carbide and Shell (2). The process is called Unipol PP and is reported to be simpler, less expensive to install, and more energy efficient than conventional polypropylene processes. The process involves a gas-phase fluidized bed process that can make a full range of polypropylene grades and can be applied to existing facilities as well as new plants. The process is being offered for licensing worldwide by Union Carbide. The process is said to use a Shell super-high-activity catalyst. Union Carbide claims that not only can existing polypropylene plants be converted to the new process but polyethylene and even polystyrene facilities can be converted essentially by replacing the reactor. The process differs from the BASF gas-phase technology in that it uses a fluidized bed rather than a stirred reactor. Carbide is converting a 200 million pound per year facility in Seadrift, Texas, which is scheduled for start-up in 1985. This first commercial-scale facility will determine the potential of this process.

Market

Since the mid-seventies overcapacity of polypropylene in the United States has resulted in a heavy discounting of prices. Consumption in 1982 was approximately 3.5 billion pounds. Capacity is in excess of 5 billion pounds. Major end uses for polypropylene are injection-molded products, which account for over one-third of the market, with about another third for fibers and about a tenth for film. Exports of polypropylene are falling from a high of about one-fifth of domestic production because of the narrowing of feedstock advantage due to oil decontrol in the United States. In addition, export customers are building polypropylene production facilities. The advantage in the future will go to those companies that have a strong feedstock position.

Internationally many countries are building and planning large petrochemical projects, most of which are based on natural gas liquids, primarily ethane. These projects will result in minimal propylene output and may ultimately result in worldwide shortages of polypropylene.

Current major U.S. producers of polypropylene are shown in Table 3-12 (8), major Western European producers in Table 3-13 (6), and major Japanese producers in Table 3-14. All Japanese producers except Mitsui, Showa Denko, and Ube use the slurry process. Mitsui and Ube use bulk polymerization, and Showa Denko uses solution polymerization.

TABLE 3-12 Major U.S. Producers of Polypropylene

Producer	Location	Estimated Capacity	
		mm lb/yr	m MT/yr
Amoco Chemicals	Chocolate Bayou, Texas	515	234
Arco	LaPorte, Texas	400	181
Eastman	Longview, Texas	140	63
Exxon	Baytown, Texas	550	249
Gulf	Cedar Bayou, Texas	400	181
Himont	Bayport, Texas	450	204
	Lake Charles, Louisiana	800	363
Northern Petro.	Morris, Illinois	200	91
Phillips	Pasadena, Texas	200	91
Rexene	Bayport, Texas	150	68
	Odessa, Texas	150	68
Shell	Norco, Louisiana	300	136
	Woodbury, New Jersey	300	136
Soltex	Deer Park, Texas	200	91
USS	LaPorte, Texas	350	159
	Neal, W. Virginia	165	75
	TOTAL	5270	2390

TABLE 3-13 Major Western European Producers of Polypropylene

Producer	Location	Estimated Capacity	
		mm lb/yr	m MT/yr
Petro. Schwechat	Schwechat, Austria	216	98
Amoco Chemicals	Geel, Belgium	298	135
Himont	Beringen, Belgium	221	100
Montefina	Feluy, Belgium	221	100
ATOCHEM	Gonfreville-L'Orcher, France	187	85
BP Chimie	Lavera, France	190	86
Shell Chimie	Berre-L'Etang, France	154	70
Soc. Norm. deMat. Plast.	Port-Jerome, France	165	75
Solvay and Cie	Sarralbe, France	198	90
BASF	Ludwigshafen, W. Germany	22	10
Hoechst	Hurth-Knapsack, W. Germany	154	70
	Kelsterbach, W. Germany	154	70
Rheinische Olefin.	Wesseling, W. Germany	243	110
Vestolen	Gelsenkirchen-Scholven, W. Germany	265	120
Montepolimeri	Brindisi, Italy	331	150
	Ferrara, Italy	165	75
	Gela, Italy	143	65
	Terni, Italy	176	80
DSM	Geleen, Netherlands	243	110
ICI	Rotterdam, Netherlands	132	60
Rotterdamse Poly.	Rotterdam, Netherlands	132	60
I/S Norpolefin	Saga, Norway	154	70

TABLE 3-13 Major Western European Producers of Polypropylene (*continued*)

Producer	Location		Estimated Capacity mm lb/yr	Estimated Capacity m MT/yr
EPSI	Sines, Portugal		110	50
Paular	Puertollano, Spain		132	60
Tarragona Quimica	Reus, Spain		88	40
ICI	Wilton, U.K.		441	200
Shell Chemicals	Carrington, U.K.		243	110
		TOTAL	5176	2349

TABLE 3-14 Major Japanese Producers of Polypropylene

Producer	Location		Estimated Capacity mm lb/yr	Estimated Capacity m MT/yr
Chisso Petrochemical	Chiba		340	154
Idemitsu Petrochemical	Chiba		350	159
Mitsubishi Petrochemical	Kashima, Yokkaichi		485	220
Mitsui Petrochemical	Chiba		265	120
Mitsui Toatsu Chemical	Ohtake, Sakai		340	154
Showa Denko	Okita		200	91
Sumitomo Chemical	Chiba, Niihama		220	100
Yokuyama Soda	Tokuyama		185	84
Tonen Petrochemical	Kawasaki		220	100
Ube Industries	Sakai		220	100
		TOTAL	2825	1282

Economics

Typical production costs for the slurry process and the gas-phase process are as follows:

Slurry Process

Capacity: 200 mm lb/yr (91 m MT/yr)
Capital cost: BLCC, $55 mm; OSBL, $20 mm; WC, $24 mm

	¢/lb	$/MT	%
Raw materials[a]	22.3	492	63
Utilities	5.2	115	15
Operating costs	1.8	40	5
Overhead costs	6.3	139	17
Cost of production	35.6	786	100
Transfer price	46.9	1034	

[a]Propylene at 20¢/lb

Gas-Phase Process

Capacity: 200 mm lb/yr (91 m MT/yr)
Capital cost: BLCC, $45 mm; OSBL, $15 mm; WC, $21 mm

	¢/lb	$/MT	%
Raw materials[a]	22.3	492	71
Utilities	2.2	49	7
Operating costs	1.5	33	5
Overhead costs	5.3	117	17
Cost of production	31.3	691	100
Transfer price	40.3	889	

[a] Propylene at 20¢/lb

OXOALCOHOLS

Oxoalcohols are generally produced by a hydroformylation reaction in which aldehydes are produced by the reaction of olefins with carbon monoxide and hydrogen. The aldehyde can then be hydrogenated to the alcohol. The alcohol and aldehyde have one carbon atom more than the feed olefin does. As an alternative procedure the aldehyde may be dimerized by an aldol condensation before hydrogenation. In this case the product will contain two more carbon atoms than twice the number in the original olefin.

In general, oxoalcohols are rated according to the number of carbon atoms in a molecule and fall into the following three main categories:

1. Solvent alcohols: mainly used in solvent applications and include propanols and butanols.
2. Plasticizer alcohols: in the range of 6 to 13 carbon atoms and used primarily in vinyl plasticizers.
3. Detergent alcohols: in the range of 11 to 20 carbon atoms and used in the production of nonionic and other surfactants.

The worldwide producers of oxoalcohols include almost all of the major petrochemical companies. Some companies produce only one or two classes, but many cover the full range.

Technology

The oxoalcohols are large-volume intermediate petrochemicals and have been manufactured for over 40 years. Until recently a catalyst system containing cobalt was used. In the last decade, however, catalyst systems containing rhodium have become commercially important for this and other carbonylation reactions.

A typical hydroformylation reaction with propylene as feed and a cobalt catalyst system is as follows:

$$CH_3CH{=}CH_2 + CO + H_2 \xrightarrow{HCo(CO)_4} CH_3CH_2CH_2CHO + \underset{\displaystyle CH_3}{\underset{|}{CH_3CHCHO}}$$

Reaction temperatures are generally 110 to 180°C, and pressures range from 200 to 350 atmospheres. The ratio of normal butanal to isobutanal in the mixed aldehyde is generally 4/1. The mixed aldehyde, when hydrogenated, yields a mixed alcohol, as shown in the following reactions:

$$CH_3CH_2CH_2CHO + H_2 \xrightarrow[10\ atm.]{Co\ or\ Ni} CH_3CH_2CH_2CH_2OH$$

$$\underset{\displaystyle CH_3}{\underset{|}{CH_3CHCHO}} + H_2 \longrightarrow \underset{\displaystyle CH_3}{\underset{|}{CH_3CHCH_2OH}}$$

Starting with propylene feed, the *n*-butanal can be dimerized in an aldol reaction to yield 2-ethylhexanol (2-EH) as follows:

$$2CH_3CH_2CHO \xrightarrow[90°C,\ 1\ atm.]{aq.\ NaOH} CH_3CH_2CH_2CH{=}\underset{\displaystyle CH_2CH_3}{\underset{|}{C}}{-}CHO$$

$$CH_3CH_2CH_2CH{=}\underset{\displaystyle CH_2CH_3}{\underset{|}{C}}{-}CHO + 2H_2 \xrightarrow[100\ atm.]{Ni} CH_3CH_2CH_2CH_2\underset{\displaystyle CH_2CH_3}{\underset{|}{C}}HCH_2OH$$

The use of a rhodium catalyst system, in general, permits a reduction of pressure to 10 to 20 atmospheres and increases the ratio of *n*-butanal to isobutanal to up to 10/1, thereby making the conditions more desirable for 2-ethylhexanol production. However, reactions with rhodium catalyst systems do generally require higher temperatures, and since the rhodium is quite expensive, care must be taken to keep the catalyst losses very low. Figure 3-6 is a block flow diagram of a typical process.

Much of the worldwide production of C_4 oxoalcohols still uses an unmodified cobalt catalyst, which has been used for many years by Ruhrchemie, BASF, PCUK, and ICI. Shell and Exxon have developed processes that use a modified cobalt catalyst, which is claimed to have higher selectivity to the normal product as well as other advantages. Recent efforts by such companies as Mitsubishi Chemical in Japan and Union Carbide in the United States have resulted in rhodium and modified rhodium catalyst systems.

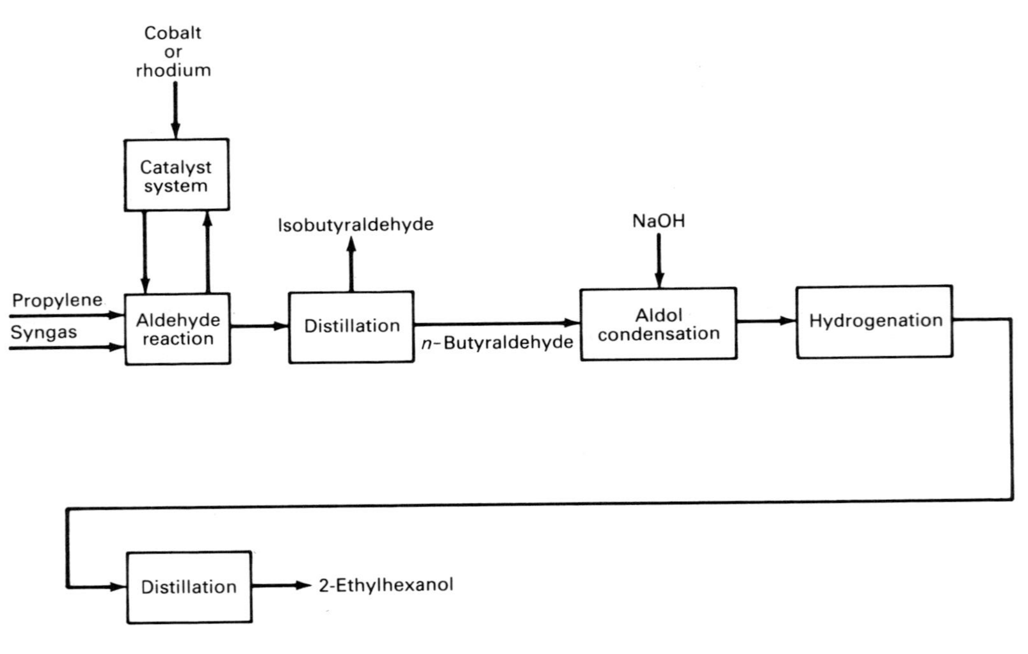

Figure 3.6 Production of 2-ethylhexanol

Before the development of the oxoalcohols process, the dominant alcohol for plasticizer use was 2-ethylhexanol, which was primarily produced from acetylene by aldol condensation. Straight-chain alcohols from coconut oil were also used to some degree. However, the oxo process permitted inexpensive propylene polymer olefins, such as heptenes, nonenes, and tetrapropylenes, to be used as feed.

Market

Solvent alcohols. In much of the world, the use of *n*-butanol in industrial coatings is the largest market for oxoalcohols. *n*-Butanol is generally the preferred solvent where slow drying at elevated temperatures is required. Slow drying is an advantage in applications such as automobile painting and wood and metal furniture finishing. In general, fewer environmental concerns are associated with *n*-butanol than with most other solvents. For many applications *n*-butanol and isobutanol are interchangeable, and demand depends on price and regional availability. In the United States about 1.5 billion pounds per year of butyraldehyde, the precursor for *n*-butanol, are produced, with about 10 percent ending up as isobutanol, the major by-product of *n*-butanol production. Regarding chemical derivatives of *n*-butanol, the two largest productions in the United States are butyl acetate and butyl methacrylate.

Plasticizer alcohols. The C_6 to C_{13} alcohols represent the largest volume of oxoalcohols produced in the United States. In the United States, as well as in Western Europe and Japan, their use is mainly in the production of flexible polyvinyl chloride as plasticizers, primarily phthalates. Dioctyl phthalate, made from 2-ethylhexanol, is probably the most common plasticizer. The 2-ethylhexanol market will therefore follow the market for PVC.

Detergent alcohols. The detergent industry is the largest consumer of C_{11} to C_{20} linear alcohols, and in the United States it accounts for about half of the demand for these alcohols. The detergent market in the developed areas of the world is expected to have a fairly low growth rate, although the use of individual products within the market could vary considerably because of different washing habits and conditions around the world, new product types, and governmental regulations. For example, a few years ago in the United States, concern over pollution problems created by nonbiodegradable detergents caused a rapid increase in the use of linear alcohol detergents. These compounds do break down under bacterial action.

Current major producers of *n*-butanol in the United States are shown in Table 3-15 (2). Current U.S. capacity exceeds 1.5 billion pounds per year (3). Consumption in 1983 reached about 700 million pounds as compared with about

TABLE 3-15 Current U.S. Producers of n-Butanol

Producer	Location	Estimated Capacity	
		mm lb/yr	m MT/yr
Badische	Freeport, Texas	100	45
Celanese	Bay City, Texas	250	113
	Bishop, Texas	175	79
Conoco (DuPont)	Westlake, Louisiana	5	2
Eastman	Longview, Texas	190	86
Shell	Deer Park, Texas	200	91
Union Carbide	Ponce, Puerto Rico	270	122
	Texas City, Texas	400	181
	TOTAL	1590	719

620 million pounds in 1982. A substantial overcapacity apparently exists (4). Therefore modest operating rates will likely be the rule for some years. There is, of course, the possibility that the product could be used as an alcohol cosolvent for consumption in the gasoline pool. Tertiary butyl alcohol is currently blended in the octane booster marketed by Arco Chemical. The percentages of *n*-butyl alcohol used for various end products are shown in Table 3-16.

Regarding 2-ethylhexanol, U.S. demand was about 460 million pounds in 1983 and close to 600 million pounds in 1984. Substantial overcapacity exists, and growth is expected to be moderately slow. Major U.S. producers of 2-ethylhexanol are shown in Table 3-17 (2). In addition to these producers, BASF has a 125 million pound per year facility at Montreal, Quebec, Canada, which provides feed for a Badische plasticizer plant in New Jersey. Capacities must be considered as flexible for these producers, since other alcohols can be produced in the same facility. The major use for the 2-ethylhexanol is in the production

TABLE 3-16 Percentage of *n*-Butyl Alcohol Used for Various End Products

End Products	%
Acrylates and methacrylate	30
Glycol ethers	22
Butyl acetate	12
Solvent	11
Plasticizers	9
Amino resins	7
Other	2
Exports	7

TABLE 3-17 Major U.S. Producers of 2-Ethylhexanol

Producer	Location	Estimated Capacity	
		mm lb/yr	m MT/yr
Badische	Freeport, Texas	55	25
Eastman	Longview, Texas	190	86
Shell	Deer Park, Texas	100	45
Tenn-USS	Pasadena, Texas	190	86
Union Carbide	Texas City, Texas	110	50
	TOTAL	645	292

of plasticizers, which takes about 58 percent of the total output. About 70 percent of this portion is used for dioctyl phthalate (2).

Table 3-18 indicates Western European production of oxo process chemicals, including alcohols, aldehydes, and acids (6). Note that the estimated capacities shown for some producers are not for alcohols only. However, in most instances most of the capacity shown is for alcohol production, primarily *n*-butyl alcohol and 2-ethylhexanol.

Major Japanese producers of oxoalcohols and their estimated capacities are shown in Table 3-19. Major products included in the capacity figures are normal and isobutanol, 2-ethylhexanol, and in a few cases heptanol and higher alcohols. Over 50 percent of the production is 2-ethylhexanol, and about 40 percent is normal and isobutanol.

TABLE 3-18 Western European Producers of Oxo Process Chemicals

Producer	Location	Estimated Capacity	
		mm lb/yr	m MT/yr
Chemie Linz	Schwechat, Austria	146	66
Norsolor	Harnes, France	287	130
Oxochimie	Lavera, France	386	175
BASF	Ludwigshafen, W. Germany	959	435
	Tarragona, Spain	146	66
Chemische Werke Huels	Marl, W. Germany	606	275
Ruhrchemie	Oberhausen, W. Germany	717	325
Montepolimeri	Priolo, Italy	88	40
Beroxo	Ornskoldsvik, Sweden	88	40
	Stenungsund, Sweden	121	55
Shell	Ellesmere Port, U.K.	66	30
	TOTAL	3610	1637

TABLE 3-19 Major Japanese Producers of Oxoalcohols

Producer	Location	Estimated Capacity	
		mm lb/yr	m MT/yr
Nippon Butanol	Yokkaichi, Kashima	180	82
Kyowa Yuka	Yokkaichi	370	168
Mitsubishi Chemical	Yokkaichi, Mizushima	360	163
Mitsubishi Petrochemical	Kashima	44	20
Chisso Chemical	Goi	100	45
Tonen Petrochemical	Kawasaki	62	28
Nissan Petrochemical	Goi	88	40
	TOTAL	1204	546

Economics

Typical production costs for a 2-ethylhexanol facility using a rhodium catalyst are as follows:

Capacity: 175 mm lb/yr (79 m MT/yr)
Capital cost: BLCC, \$38 mm; OSBL, \$20 mm; WC, \$17 mm

	¢/lb	\$/MT	%
Raw materials[a]	24.4	538	71
Utilities	3.3	73	10
Operating costs	1.3	29	4
Overhead costs	5.3	117	15
Cost of production	34.3	757	100
By-product credit[b]	(5.0)	(110)	
Net cost of production	29.3	647	
Transfer price	39.2	864	

[a] Propylene at 20¢/lb
[b] Isobutyraldehyde at 26¢/lb

1,4-BUTANEDIOL

1,4-Butanediol is a raw material for many products, including polybutylene terephthalate (PBT), polyurethanes, and tetrohydrofuran (THF). PBT is a rapidly growing engineering thermoplastic, and considerable effort has been expended in developing other routes to the butanediol. Most of the butanediol at present is commercially produced from acetylene by Reppe chemistry. A major factor in the relatively high cost of the butanediol is the cost of the acetylene. Much of the recent technical literature has been concerned with propylene as a potential feedstock for the 1,4-butanediol, and it has therefore been included in this section on propylene derivatives.

Technology

Acetylene-based production. In this process acetylene is condensed with formaldehyde using a copper catalyst to produce butynediol. This material is then hydrogenated, generally with a nickel catalyst, to produce the butanediol. The reactions are as follows:

$$HC{\equiv}CH + 2HCHO \xrightarrow{Cu} HOCH_2C{\equiv}CCH_2OH$$

$$HOCH_2C{\equiv}CCH_2OH + 2H_2 \xrightarrow{Ni} HO(CH_2)_4OH$$

A block flow diagram of this process is shown in Figure 3-7.

The following feedstocks are being considered as substitutes for the expensive acetylene:

1. Propylene: via allyl acetate, acrolein, or allyl alcohol.
2. Butadiene: via 1,4-diacetoxybutene-2 or butadiene peroxide.
3. Normal butane or benzene: via maleic anhydride.

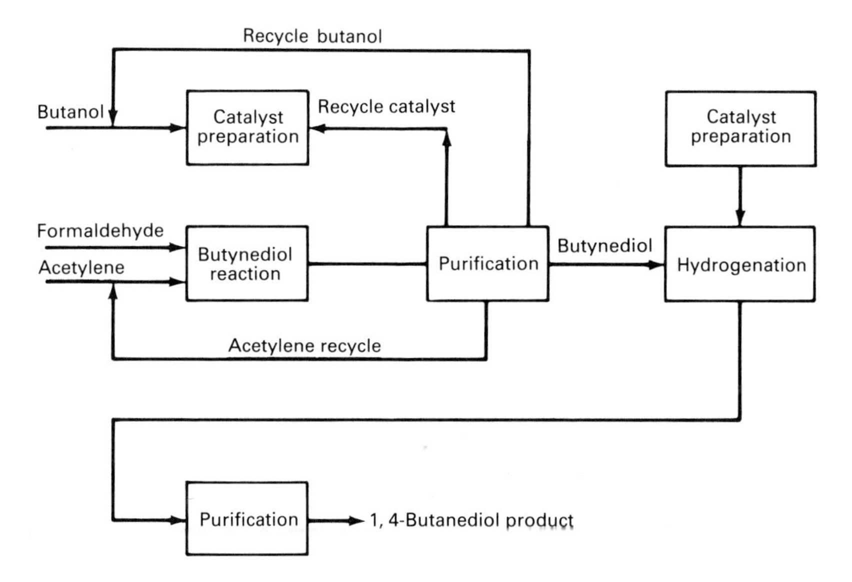

Figure 3.7 Production of 1,4-butanediol via acetylene and formaldehyde

Propylene-based production. With propylene used as the original raw material, basically three different routes are being considered. General Electric technology involves, as an initial step, the vapor-phase acetoxylation of propylene to allyl acetate. Following this step, the allyl acetate is hydrocarbonylated to 4-acetoxybutanol and subsequently hydrolyzed to the 1,4-butanediol as follows:

$$CH_2CHCH_2OAc + CO + H_2 \xrightarrow{Co_2(CO)_8} HO(CH_2)_4OAc$$

$$HO(CH_2)_4OAc + H_2O \xrightarrow{acid} HO(CH_2)_4OH + HOAc$$

The hydrocarbonylation uses a cobalt catalyst at 125°C and 3000 psi. The reaction is exothermic. One disadvantage of this process is the production of appreciable amounts of mixed diols—about a third of a pound of mixed diols per pound of the 1,4-butanediol. Kuraray has developed a process from propylene via allyl alcohol (3). The process appears to be strongly dependent on the cost of producing the allyl alcohol. Allyl alcohol can be produced by the acid-catalyzed isomerization of propylene oxide, the reduction of the acrolein, or the hydrolysis of allyl acetate. The allyl alcohol is hydroformylated to hydroxy-butyraldehyde using a rhodium catalyst, which is followed by hydrogenation to produce the butanediol:

$$CH_2{=}CHCH_2OH + H_2 + CO \xrightarrow{Rh} OHCCH_2CH_2CH_2OH$$

$$OHCCH_2CH_2CH_2OH + H_2 \xrightarrow{Ni} HO(CH_2)_4OH$$

Butadiene-based production. Mitsubishi Chemical Industries has developed a three-stage process from butadiene (4). The first stage involves the liquid- or gas-phase acetoxylation of butadiene to 1,4-diacetoxy-2-butene over a supported palladium catalyst. Following this reaction, hydrogenation and hydrolysis produce the 1,4-butanediol. The reactions follow:

$$CH_2{=}CHCH{=}CH_2 + 2HOAc \longrightarrow AcOCH_2CH{=}CHCH_2OAc$$

$$AcCH_2CH{=}CHCH_2OAc + H_2 \longrightarrow AcOCH_2CH_2CH_2CH_2OAc$$

$$AcOCH_2CH_2CH_2CH_2OAc + H_2O \longrightarrow HO(CH_2)_4OH + 2HOAc$$

Continuing work is under way to improve selectivity in the initial acetoxylation reaction. Another possible route starting with butadiene is based on butadiene peroxide as an intermediate. Mitsubishi has constructed a 330 million pound per year facility at Yokkaichi, Japan, based on butadiene and acetic acid.

Note that acetylene is a scarce and expensive feedstock in Japan, which would affect the economics vis-a-vis the Reppe chemistry route in a way that might not be valid in Europe or the United States.

Butane- or benzene-based production. The first step in this technology involves the production of maleic anhydride. Up to a few years ago, the preferred feedstock was benzene, but more recently the switch has been to the use of normal butane or even normal butene. Once the maleic anhydride has been produced, hydrogenation can produce tetrahydrofuran and 1,4-butanediol via succinic anhydride and butyrolactone as follows:

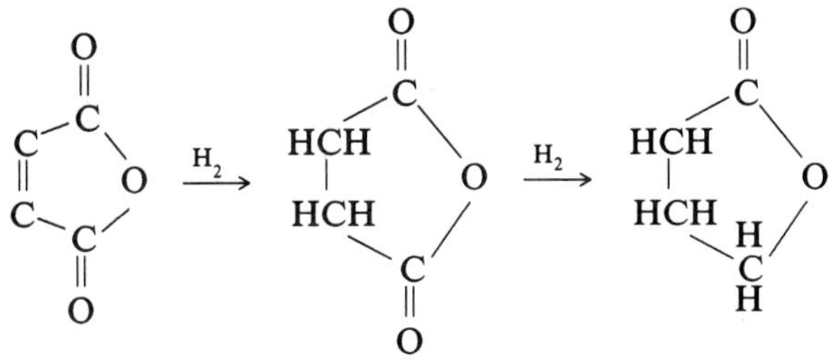

$$\text{MAN} \xrightarrow{H_2} \text{Succinic Anhydride} \xrightarrow{H_2} \text{Butyrolactone}$$

MAN Succinic Anhydride Butyrolactone

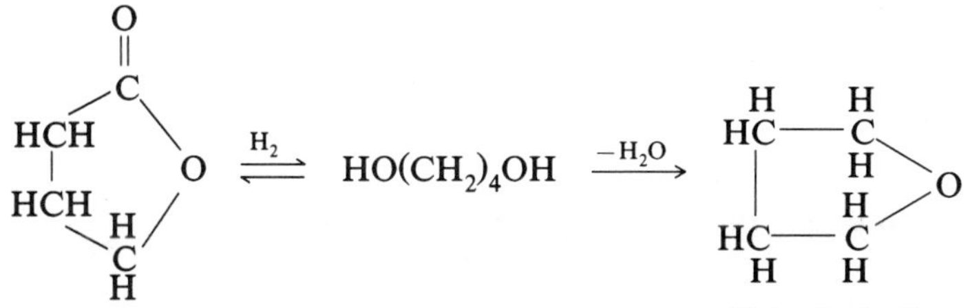

$$\text{Butyrolactone} \underset{}{\overset{H_2}{\rightleftharpoons}} HO(CH_2)_4OH \xrightarrow{-H_2O} \text{Tetrahydrofuran}$$

Tetrahydrofuran

The initial reaction step produces succinic anhydride, and further hydrogenation produces the butyrolactone. Hydrogenation of the butyrolactone to 1,4-butanediol is virtually quantitative. The principal by-product is the tetrohydrofuran.

Market

The market for 1,4-butanediol is essentially captive. Perhaps a third of the product enters the merchant market directly, mostly for use in polyurethanes and polybutylene terephthalate. Furthermore perhaps half of the captive use is

TABLE 3-20 U.S. Producers of 1,4-Butanediol

Producer	Location	Estimated Capacity	
		mm lb/yr	m MT/yr
BASF Wyandotte	Geismar, Louisiana	55	25
GAF	Calvert City, Kentucky	55	25
	Texas City, Texas	55	25
DuPont	LaPorte, Texas	210	95
	TOTAL	375	170

in the manufacture of tetrahydrofuran, a widely used solvent. The most important growth area for the 1,4-butanediol is generally expected to be PBT. This end use is still fairly small, but its growth is considerable. For the near future the main outlet will probably be tetrahydrofuran. One point to keep in mind is that furfural is a very viable feedstock for the production of tetrahydrofuran, and the production of this material as a by-product of cellulose-based ethanol facilities could be considerable if there were substantial use of ethanol as a fuel or octane enhancer in the United States.

In the United States DuPont, GAF, and BASF-Wyandotte are the producers of 1,4-butanediol, all produced from acetylene via Reppe chemistry. Locations and estimated capacities are shown in Table 3-20(5). In West Germany BASF and GAF-Huels produce 1,4-butanediol from acetylene. They are the only producers in Western Europe (6). The BASF facility is in Ludwigshafen and has a capacity of 110,000 metric tons per year (243 mm lb/yr), and the GAF-Huels facility is in Marl and has a capacity of 30,000 metric tons per year (65 mm lb/yr). At present only Japan uses other feedstocks commercially, because of the high cost and scarcity of acetylene, as discussed earlier. Japan has traditionally exported butadiene, since it has had an excess and the material was usually less expensive in Japan than in the United States. However, with the decrease in ethylene production, the amount of available butadiene decreased and prices in the two countries are currently about equal. Toyo Soda starts with butadiene via dichlorobutene to produce a small amount of 1,4-butanediol, and Mitsubishi Petrochemical and Mitsubishi Chemical Industries produce a small amount from maleic anhydride. Companies involved in the development of new processes based on butadiene via acetoxylation are Mitsubishi Chemical, BASF, Japan Synthetic Rubber, Toyo Soda, Sumitomo, Asahi, and Mitsui Toatsu. Toyo Soda and Mitsubishi Petrochemical are also investigating processes based on butadiene via butadiene peroxide.

Economics

Typical production cost for 1,4-butanediol from acetylene via Reppe chemistry are shown in the following table. Note that a large variation occurs in the cost of acetylene, since many sources are used and the accounting procedures of the various companies producing it vary widely. Studies made by the author indicate that the capital cost for a facility using butadiene feedstock is higher than that for a facility using Reppe chemistry. For facilities using propylene and maleic anhydride, feedstock cost is lower.

Capacity: 66 mm lb/yr (30 m MT/yr)
Capital cost: BLCC, $35 mm; OSBL, $13 mm; WC, $15 mm

	¢/lb	$/MT	%
Raw materials[a]	42.7	942	64
Utilities	8.0	176	12
Operating costs	3.2	71	5
Overhead costs	13.2	291	19
Cost of production	67.1	1480	100
Transfer price	88.9	1960	

[a] Acetylene at 50¢/lb

REFERENCES

Propylene Oxide and Propylene Glycol

1. List, H. L. "Propylene Oxide," *International Petrochemical Developments,* Vol. 2, No. 6, March 15, 1981.
2. *1984 Directory of Chemical Producers—United States,* SRI International.
3. "Chemical Profiles," *Chemical Marketing Reporter,* February 27, 1984.
4. "Petrochemical Handbook '83," *Hydrocarbon Processing,* November 1983.
5. *1984 Directory of Chemical Producers—Western Europe,* SRI International.
6. *Kirk-Othmer Encyclopedia of Chemical Technology,* Third Edition, Vol. 11. New York: Wiley, 1984, pp. 933–56.
7. Ibid., Vol. 19, 246–74.
8. "Chemical Profiles," *Chemical Marketing Reporter,* February 20, 1984.
9. *JCW Chemicals Guide 82/83.* Japan: The Chemical Daily Co. Ltd., March 1982.
10. *Chemical Industry Yearbook,* Second Edition. Surrey, England: Industrial Press, 1984.

Acrylonitrile

1. List, H. L. "Acrylonitrile—A Status Report," *International Petrochemical Developments,* Vol. 2, No. 20, October 15, 1981.
2. "Chemical Profile," *Chemical Marketing Reporter,* April 16, 1984.
3. *1984 Directory of Chemical Producers—United States,* SRI International.
4. "Petrochemical Handbook '83," *Hydrocarbon Processing,* November 1983.
5. *1984 Directory of Chemical Producers—Western Europe,* SRI International.
6. *Kirk-Othmer Encyclopedia of Chemical Technology,* Third Edition, Vol. 1. New York: Wiley, 1984, pp. 414–26.
7. *JCW Chemicals Guide 82/83.* Japan: The Chemical Daily Co. Ltd., March 1982.
8. *Chemical Industry Yearbook,* Second Edition. Surrey, England: Industrial Press, 1984.

Acrylic Fibers

1. *1984 Directory of Chemical Producers—United States,* SRI International.
2. *1984 Directory of Chemical Producers—Western Europe,* SRI International.
3. "Petrochemical Handbook '83," *Hydrocarbon Processing,* November 1983.

Polypropylene

1. List, H. L. "Polypropylene—A Status Report," *International Petrochemical Developments,* Vol. 2, No. 8, April 15, 1981.
2. "Polypropylene: New Process is Simpler, Cheaper," *Chemical and Engineering News,* November 14, 1983, pp. 6–7.
3. Mannon, J. H. "Polypropylene: Battle for the High Ground," *Chemical Business,* May 1984, pp. 19–23.
4. "PP Process Cuts Energy Needs 75%," *Oil and Gas Journal,* January 9, 1984, p. 92.
5. "Petrochemical Handbook '83," *Hydrocarbon Processing,* November 1983.
6. *1984 Directory of Chemical Producers—Western Europe,* SRI International.
7. *Kirk-Othmer Encyclopedia of Chemical Technology,* Third Edition, Vol. 16. New York: Wiley, 1984, pp. 453–70.
8. *1984 Directory of Chemical Producers—United States,* SRI International.
9. *JCW Chemicals Guide 82/83.* Japan: The Chemical Daily Co. Ltd., March 1982.
10. "Materials '85," *Modern Plastics,* January 1985, pp. 61–71.
11. DiDrusco, G. and R. Rinaldi. "Polypropylene—Process Selection Criteria," *Hydrocarbon Processing,* November 1984, pp. 113–17.
12. *Chemical Industry Yearbook,* Second Edition. Surrey, England: Industrial Press, 1984.

Oxoalcohols

1. List, H. L. "Oxo-Alcohols—A Status Report," *International Petrochemical Developments,* Vol. 2, No. 12, June 15, 1981.
2. "Chemical Profiles," *Chemical Marketing Reporter,* December 7, 1981.
3. "Butanol Consumption May Climb to 700 Million Pounds This Year, but Overcapacity Clouds Future," *Chemical Marketing Reporter,* June 23, 1983.
4. "Butanol Makers 1983 Rise Seems Likely to Plateau in Tandam with Housing, Autos," *Chemical Marketing Reporter,* February 27, 1984.
5. "Petrochemical Handbook '83," *Hydrocarbon Processing,* November 1983.
6. *1984 Directory of Chemical Producers—Western Europe,* SRI International.
7. *Kirk-Othmer Encyclopedia of Chemical Technology,* Third Edition, Vol. 16. New York: Wiley, 1984, pp. 637–53.
8. *1984 Directory of Chemical Producers—United States*, SRI International.

1,4-Butanediol

1. Brownstein, A. M. and H. L. List. "Which Route to 1,4-Butanediol," *Hydrocarbon Processing,* September 1977, pp. 159–62.
2. List, H. L. "Which Feedstock for 1,4-Butanediol," *International Petrochemical Developments,* Vol. 2, No. 2, January 15, 1981.

3. Tamura, M. and S. Kumano. "New Process for 1,4-Butanediol via Allyl Alcohol," *Chemical Economy and Engineering Review,* Vol. 12, No. 9 (No. 141), September 1980, pp. 32–36.
4. Tanabe, Y. "New Route to 1,4-Butanediol and Tetrahydrofuran," *Hydrocarbon Processing,* September 1981, pp. 187–90.
5. *1984 Directory of Chemical Producers—United States,* SRI International.
6. *1984 Directory of Chemical Producers—Western Europe,* SRI International.
7. *Kirk-Othmer Encyclopedia of Chemical Technology,* Third Edition, Vol. 11. New York: Wiley, 1984, pp. 956–62.
8. *JCW Chemicals Guide 82/83.* Japan: The Chemical Daily Co. Ltd., March 1982.

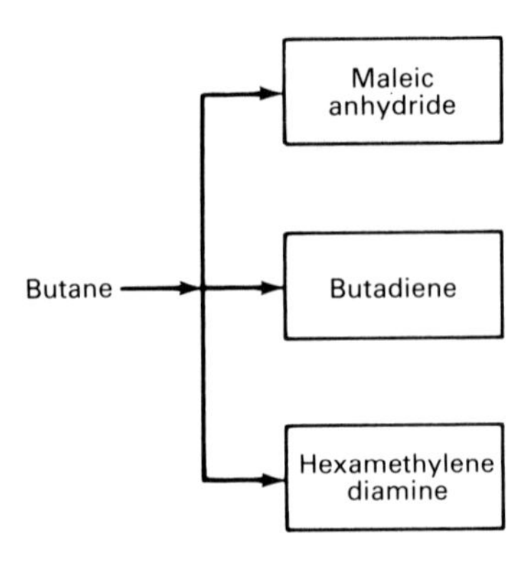

Butane derivatives

CHAPTER FOUR

Butane Derivatives

MALEIC ANHYDRIDE (MAN)

Over the past few years, the market for maleic anhydride has been undergoing disruptions caused largely by fluctuations in feedstock preference between butane and benzene. Demand in the United States was about 260 million pounds in 1982, rose slightly in 1983, and is projected to be in excess of 300 million pounds by 1987 (1). More than half of the material is used in the production of unsaturated polyester resins, which are used primarily in the transportation and construction industries. Butane has become the feedstock of choice.

Technology

Up to a few years ago, benzene was the primary feedstock for the production of maleic anhydride; but because of the rapid rise in the price of aromatics, the change to C_4 hydrocarbons has grown markedly. The early work in the United States was initiated by DuPont in the mid-fifties. The patent literature describes the use of a modified cobalt molybdate catalyst for the oxidation of normal butane (3,4). A decade later Princeton Chemical Research described the use of catalyst systems containing vanadium and phosphorous (5). The earliest commercial use of *n*-butane as a feedstock for maleic anhydride is believed to have been by Monsanto, who used a vanadium phosphorous catalyst modified by iron. Reaction temperatures vary and are primarily dependent on the catalyst system used.

Regardless of which feedstock is used, the design of the plant is similar and consists basically of three major sections: reaction, recovery, and purification. Generally multiple reactors are used. They are usually tubular, although fluid bed reactors are being developed (6). The reactor tubes are packed with catalyst, and the heat removal medium is most often circulating molten salt. Heats of reaction are large. Reaction temperatures are maintained in the range of 350 to 500°C, depending on the catalyst system used. The temperature and other reaction variables, such as air rate, pressure, and feed concentration, are adjusted to balance catalyst life and performance. The reactor effluent passes through heat recovery equipment and coolers, in which some of the crude MAN is condensed and subsequently refined. The noncondensed material is sent to a scrubber, where a maleic acid solution recovers most of the remaining maleic anhydride. The tail gases are burned or vented. The maleic acid is then sent to evaporators for concentration and dehydration. The MAN recovered from the scrubber is combined with the crude MAN from the condenser for final purification.

The primary reactions are as follows:

Benzene Feed

$$C_6H_6 + \tfrac{9}{2}O_2 \longrightarrow \text{HC(=O)C–CH=CH–C(=O)O (maleic anhydride)} + 2CO_2 + 2H_2O$$

$$C_6H_6 + \tfrac{15}{2}O_2 \longrightarrow 6CO_2 + 5H_2O$$

Butane Feed

$$C_4H_{10} + \tfrac{7}{2}O_2 \longrightarrow \text{(maleic anhydride)} + 4H_2O$$

$$C_4H_{10} + \tfrac{13}{2}O_2 \longrightarrow 4CO_2 + 5H_2O$$

Examination of the main reaction for the two feeds indicates a theoretical advantage for the butane feedstock. For the butane a four-carbon feedstock is converted to a four-carbon product, whereas for the benzene a six-carbon feedstock is converted to a four-carbon product.

Since the equipment is essentially the same for both feedstock, it is possible to switch from one feed to another, and some manufacturers reportedly are considering this option to take advantage of variations of feedstock costs. In general the use of butane in an existing plant that is based on benzene will lower the productivity because of the somewhat higher residence time required with the butane feed. Another related factor favoring the use of butane feedstock in the United States is the stringent benzene emission control standards proposed by the Environmental Protection Agency. A block flow diagram for the process is shown in Figure 4-1.

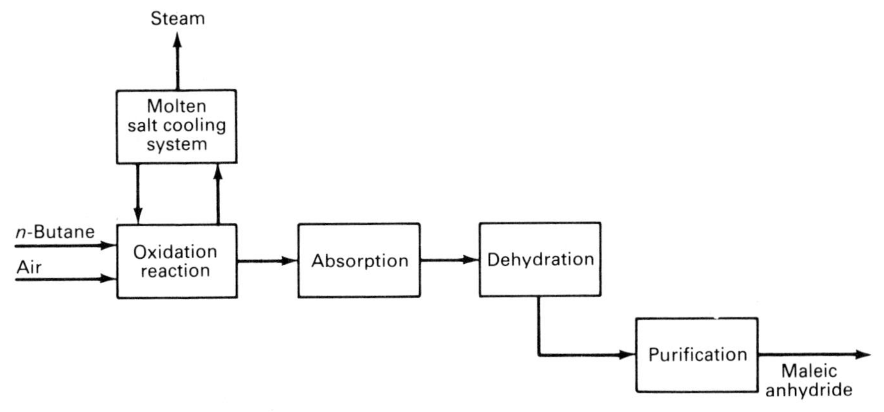

Figure 4.1 Production of MAN—typical air oxidation process

Market

Currently five major producers of maleic anhydride are in the United States, down from eight a few years ago. Reichhold, Tenneco, and Koppers have dropped out of the market. Koppers produced its product as a by-product of phthalic anhydride production, and Reichhold and Tenneco had facilities based on benzene feed. Current U.S. producers are shown in Table 4-1 (1,7).

U.S. producers use predominately *n*-butane feed. The Monsanto facility at St. Louis has converted some capacity to *n*-butane from benzene and has

TABLE 4-1 Current U.S. Producers of Maleic Anhydride

		Estimated Capacity	
Producer	Location	mm lb/yr	m MT/yr
Amoco	Joliet, Illinois	60	27
Ashland	Neal, W. Virginia	60	27
Denka	Houston, Texas	50	23
Monsanto	Pensacola, Florida	130	59
	St. Louis, Missouri	40	18
USS	Neville Island, Pennsylvania	55	25
	TOTAL	395	179

shut down about 70 million pounds per year of capacity based on benzene. The USS facility is also undergoing conversion of reactors to *n*-butane from benzene. Outside of the United States, in addition to benzene and *n*-butane feedstocks, some maleic anhydride is produced in West Germany and Japan by the oxidation of *n*-butenes. Major producers in Western Europe are shown in Table 4-2 (11).

TABLE 4-2 Major Western European Producers of Maleic Anhydride

Producer	Location	Estimated Capacity		Feedstock
		mm lb/yr	m MT/yr	
Chemie Linz	Linz, Austria	26	12	Benzene
UCB-Ftal	Oostende, Belgium	7	3	PAN by-product
CdF Chimie	Drocourt, France	31	14	Benzene
	Villers-Saint-Paul, France	31	14	Benzene
BASF	Ludwigshafen, W. Germany	7	3	PAN by-product
Bayer	Krefeld, W. Germany	22	10	Butane
	Leverkusen, W. Germany	4	2	PAN by-product
Chemische Werke Huels	Bottrop, W. Germany	110	50	Butane
		7	3	PAN by-product
Deutsche Texaco	Moers, W. Germany	18	8	Benzene
Alusuisse	Scanzorosciale, Italy	60	27	Benzene
		13	6	Butane
Montepolimeri	Mantova, Italy	11	5	Benzene
Terni Ind. Chem.	Porto-Torres, Italy	44	20	Benzene
CEPSA	Algeciras, Spain	22	10	Benzene
Monsanto	Newport, Wales, U.K.	55	25	Butane
	TOTAL	468	212	

Economics

Typical production costs for both butane and benzene feed are as follows:

Butane Feed

Capacity: 60 mm lb/yr (27 m MT/yr)
Capital cost: BLCC, \$38 mm; OSBL, \$12 mm; WC, \$7 mm

	¢/lb	\$/MT	%
Raw materials[a]	14.8	326	43
Utilities	3.9	86	11
Operating costs	3.3	73	10
Overhead costs	12.2	269	36
Cost of production	34.2	754	100
By-product credit (steam)	(4.1)	(90)	
Net cost of production	30.1	664	
Transfer price	55.1	1215	

[a] Butane at 11¢/lb

Benzene Feed

Capacity: 60 mm lb/yr (27 m MT/yr)
Capital cost: BLCC, $34 mm; OSBL, $11 mm; WC, $7 mm

	¢/lb	$/MT	%
Raw materials[a]	20.9	461	55
Utilities	3.3	73	9
Operating costs	2.9	64	8
Overhead costs	11.0	243	28
Cost of production	38.1	841	100
By-product credit (steam)	(4.5)	(99)	
Net cost of production	33.6	742	
Transfer price	56.1	1237	

[a] Benzene at 17¢/lb

BUTADIENE

Butadiene is obtained as a by-product in the production of ethylene and propylene by the steam cracking of hydrocarbons. A secondary source is by the dehydrogenation of butane and butylenes. Uses of the diolefin are largely in the manufacture of synthetic elastomers and as an intermediate in the synthesis of many petrochemicals.

Technology

The yields of butadiene generally increase as the molecular weight of the feedstock increases, but the increase only becomes significant when naphtha or heavier feeds are used. In the cracking furnace, hydrocarbons are pyrolyzed in the presence of steam at temperatures of about 815°C. The temperature varies somewhat depending on the feedstock used, the type of furnace, and the severity of cracking.

Ethylene producers do not generally have substantial control over the by-product butadiene production. In essence they take what they can get. The major feedstocks used worldwide for ethylene production are as follows:

1. Gas (ethane, propane, butane)
2. Naphtha
3. Heavy distillates (gas oil)

In Europe and Japan feedstock is primarily naphtha, and thus substantial quantities of butadiene are produced. Although future trends will be the increased use of gas (from North Sea deposits in the case of Europe) and gas oil, naphtha will still represent a large portion of the feedstocks for the next decade. In the United States most ethylene is produced by gas cracking, which is the result of the low price and availability of natural gas liquids. Because of the use of this feedstock by-product production of butadiene represents only a small portion of the total butadiene demand. The trend in the United States is toward increasing the use of naphtha as a feedstock, resulting in expansion in butadiene production from U.S. cracking operations.

The income from a steam cracking operation is largely due to the sale of the ethylene and propylene produced. However, coproduct credits are important and become a significant portion of the income when naphtha or heavier feedstocks are used. The butadiene stream is extracted and sold at chemical value. In the United States butadiene from steam cracking operations is insufficient to meet demand, and its value is usually dependent on the cost of dehydrogenating butenes and butane. In Europe more butadiene is produced as a by-product of ethylene manufacture than is required by local demand, and much of the excess is exported to the United States. Japan changed from an exporter to an importer of butadiene in the early eighties.

As discussed previously, butadiene in the United States is also produced from *n*-butane by dehydrogenation. The reaction probably occurs in two steps: first, conversion of the butane to butene; then, further conversion to butadiene:

$$C_4H_{10} \longrightarrow C_4H_8 + H_2$$
$$C_4H_8 \longrightarrow C_4H_6 + H_2$$

Both reactions are endothermic and are favored by high temperature and low pressure. The major technology at present is the Houdry process, a one-step process without intermediate separation of the butenes. A chromia-alumina catalyst is used, and temperature is about 650°C and pressure 2.5 to 5 psia. The process is cyclic and generally uses a minimum of three reactors for continuous operation. A block flowsheet of the process is shown in Figure 4-2.

As can be seen from the second reaction, butadiene can also be produced from *n*-butenes by dehydrogenation. Several alternative processes to the Houdry are possible for this reaction, including the Dow-B process, which claims higher selectivity to butadienes. This process does, however, require high steam usage and therefore substantially higher energy consumption.

In either case the increased use of naphtha feedstock in the United States will, in the long term, result in greater production of butadiene in the eighties, thereby reducing demand for butadiene from dehydrogenation units as well as from European imports. However, for the near term, production from butane is likely to continue because the operating rates for naphtha-based ethylene plants will be low. These low operating rates are a result of the continued availability of light feedstocks in the United States as well as the start-up of new ethane-based ethylene units in Canada, Mexico, and the Far East.

The following table indicates the approximate weight ratio of butadiene to ethylene for various steam cracker feedstocks:

Feedstock	Butylene/Ethylene Weight Ratio
Ethane	0.01
Propane	0.04
n-Butane	0.04
Light naphtha	0.15
Full range naphtha	0.15
Gas oil	0.19

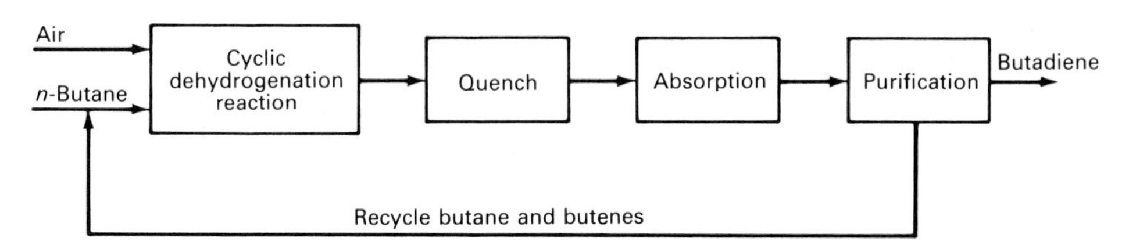

Figure 4.2 Production of butadiene by dehydrogenation of butane

Market

In the United States most of the butadiene is used in the manufacture of synthetic elastomers, primarily styrene-butadiene rubber (SBR), polybutadiene rubber (PBR), neoprene, and nitrile butadiene rubber (NBR). Most of the remainder of the butadiene is used in the production of acrylonitrile-butadiene-styrene (ABS), hexamethylene diamine (HMDA), and styrene-butadiene resins (SB).

Major U.S. butadiene producers and their estimated capacities are shown in Table 4-3. Note that these operations can change rather rapidly and the information shown in this table should be considered approximate (3).

In 1981 U.S. consumption of butadiene was close to 4 billion pounds, of which $\frac{1}{2}$ billion was imported. At that time about a third is estimated to have come from dehydrogenation and the remainder as a coproduct of ethylene production. Production was then running at about 70 percent of capacity. Many producers since that time have shut down operations or have sold their facilities.

Butadiene production in Western Europe has been curtailed because of the low operating rates of the olefins plants, but it is still exporting almost a billion pounds per year. Major producers of butadiene in Western Europe are shown in Table 4-4 (2).

TABLE 4-3 Major U.S. Producers of Butadiene

Producer	Location	Estimated Capacity		Remarks
		mm lb/yr	m MT/yr	
Amoco	Chocolate Bayou, Texas	180	82	Ethylene by-product
Arco	Channelview, Texas	450	204	Ethylene by-product Butane dehydrogenation
Dow	Freeport, Texas	85	39	Ethylene by-product
DuPont	Chocolate Bayou, Texas	135	61	Ethylene by-product
El Paso	Corpus Christi, Texas	220	100	Ethylene by-product
Exxon	Baton Rouge, Louisiana	310	141	Ethylene by-product
	Baytown, Texas	240	109	Ethylene by-product
Mobil	Beaumont, Texas	80	36	Ethylene by-product
Shell	Deer Park, Texas	500	227	Ethylene by-product
	Norco, Louisiana	500	227	Ethylene by-product
Tenneco	Houston, Texas	800	363	Butane dehydrogenation
Texaco	Port Neches, Texas	500	227	Ethylene by-product
Union Carbide	Seadrift, Texas	33	15	Ethylene by-product
	Texas City, Texas	55	25	Ethylene by-product
	Taft, Louisiana	75	34	Ethylene by-product
	Penuelas, Puerto Rico	75	34	Ethylene by-product
	TOTAL	4238	1924	

TABLE 4-4 Major Western European Producers of Butadiene

Producer	Location	Estimated Capacity	
		mm lb/yr	m MT/yr
OMV	Schwechat, Austria	110	50
Neste Oy	Kullo, Finland	44	20
Vapocraqueur	Feyzin, France	110	50
ATOCHEM	Gonfreville-L'Orcher, France	132	60
Esso Chemie	Notre-Dame-de-Gravenchon, France	154	70
Naphtachimie	Lavera, France	221	100
Shell Chimie	Berre-L'Etang, France	165	75
Bunawerke Huls	Marl, W. Germany	265	120
EC Erdolchemie	Koln, W. Germany	276	125

TABLE 4-4 Major Western European Producers of Butadiene (*continued*)

Producer	Location	Estimated Capacity mm lb/yr	m MT/yr
Esso Chemie	Koln, W. Germany	51	23
Rheinische Olefinwerke	Wesseling, W. Germany	386	175
Ruhr Oel	Gelsenkirche, W. Germany	66	30
Enichem Polimeri	Porto-Torres, Italy	110	50
	Ravenna, Italy	254	115
Riveda Srl	Brindisi, Italy	276	125
Dow Chemical	Terneuzen, Netherlands	353	160
DSM	Geleen, Netherlands	221	100
Shell Chemie	Moerdijk, Netherlands	221	100
CNP	Sines, Portugal	93	42
CALATRAVA	Puertollano, Spain	71	32
	Tarragona, Spain	221	100
BP Chemicals	Grangemouth, U.K.	110	50
	Port Talbot, Wales, U.K.	132	60
	Wilton, U.K.	198	90
Esso Chemical	Fawley, U.K.	154	70
ICI	Wilton, U.K.	209	95
Shell Chemicals	Carrington, U.K.	66	30
	TOTAL	4669	2117

Because a good portion of butadiene is used in the production of SBR, the demand for butadiene is closely tied to the fortunes of the tire industry, since SBR is the most important synthetic rubber for tires. About two-thirds of the SBR production is used in tires; therefore it is very dependent on the fortunes of the automobile industry. In addition, the trend toward radial tires at the expense of belted-bias tires has hurt the industry. Radial tires contain about one-third natural rubber, as opposed to half that much for the belted-bias tires.

In the United States and Europe, butadiene is primarily sold on the merchant market, although a few producers, such as Shell, Huls, and ANIC, are also significant producers of synthetic rubber. In the United States some rubber companies are also actively involved in butane or butene dehydrogenation facilities. In Japan, however, over 80 percent of the butadiene produced is used captively. Japanese producers include Mitsubishi Chemical Industries, Mitsui Toatsu Chemicals, and Takeda Chemical Industries.

Economics

Typical production costs for butadiene via extraction from raw C_4's are as follows:

Capacity: 200 mm lb/yr (91 m MT/yr)
Capital cost: BLCC, \$44 mm; OSBL, \$22 mm; WC, \$16 mm

	¢/lb	\$/MT	%
Raw materials[a]	36.2	798	81
Utilities	2.3	51	5
Operating costs	1.2	26	3
Overhead costs	5.2	115	11
Cost of production	44.9	990	100
By-product[b]	(20.0)	(441)	
Net cost of production	24.9	549	
Transfer price	34.8	767	

[a] Raw C_4's at 16¢/lb
[b] C_4 raffinate at 16¢/lb

HEXAMETHYLENEDIAMINE

Hexamethylenediamine (HMDA) is an important nylon intermediate and, as such, essentially follows the fortunes of nylon fibers and plastics. Essentially all nylon fibers and plastics are made in two basic ways. Nylon 6/6, which accounts for most of the U.S. production, is the condensation product of HMDA and adipic acid. Nylon 6, which is produced from caprolactam, is the major nylon produced outside the United States. The difference in production locations is undoubtedly due to the fact that Nylon 6/6 was developed in the United States by DuPont in the late thirties and Nylon 6 by BASF in Germany at about the same time.

Technology

HMDA has historically been made from adipic acid. At temperatures of about 300°C, adipic acid reacts with ammonia to produce adiponitrile in accordance with the following reaction:

$$HOOC(CH_2)_4COOH + 2NH_3 \longrightarrow NC(CH_2)_4CN$$

In the presence of phosphoric acid, yields are generally in excess of 90 percent. The adiponitrile is then hydrogenated to hexamethylenediamine as follows:

$$NC(CH_2)_4CN + 4H_2 \longrightarrow H_2N(CH_2)_4NH_2$$

New technology has largely displaced this process.

HMDA is essentially a captive product and is concentrated in the hands of those producers who manufacture the nylon. DuPont and Monsanto together account for almost two-thirds of the U.S. nylon fiber market. DuPont produces its HMDA by the hydrocyanation of butadiene. The initial step is the addition of HCN to butadiene, employing a copper chromite catalyst. Both linear and branched cyanobutenes in a ratio of about 7/1 are produced. The linear reaction is as follows:

$$CH_2{=}CHCH{=}CH_2 + HCN \longrightarrow CH_2CH{=}CHCH_2CN \quad \text{or} \quad CH_2{=}CHCH_2CH_2CN$$

Conversion is virtually quantitative. Excess butadiene is avoided so as to minimize its dimerization. The linear products are separated and further reacted with additional HCN to produce adiponitrile. Conversion is about 98 percent, and yield of adiponitrile is over 90 percent as follows:

$$CH_2CH{=}CHCH_2CN + HCN \longrightarrow NC(CH_2)_4CN$$

The catalyst system typically includes a solution of stannous chloride in tetrohydrofuran. The final step in the DuPont technology involves the hydrogenation of adiponitrile to HMDA in the liquid phase, employing a Raney nickel catalyst system as follows:

$$NC(CH_2)_4 + 4H_2 \longrightarrow H_2N(CH_2)_6NH_2$$

Figure 4-3 represents a block flow diagram for the process.

Monsanto's technology involves the electrohydrodimerization of acrylonitrile. Authorities claim that adiponitrile can be obtained from acrylonitrile in over 90 percent yield at 50 percent conversion by electrolysis of the aqueous solution of acrylonitrile. A typical cell has a mercury or lead cathode, a silver alloy anode, and an ion exchange membrane to separate the anolyte and catholyte sections. A quaternary ammonium salt typically serves as a catholyte, and the

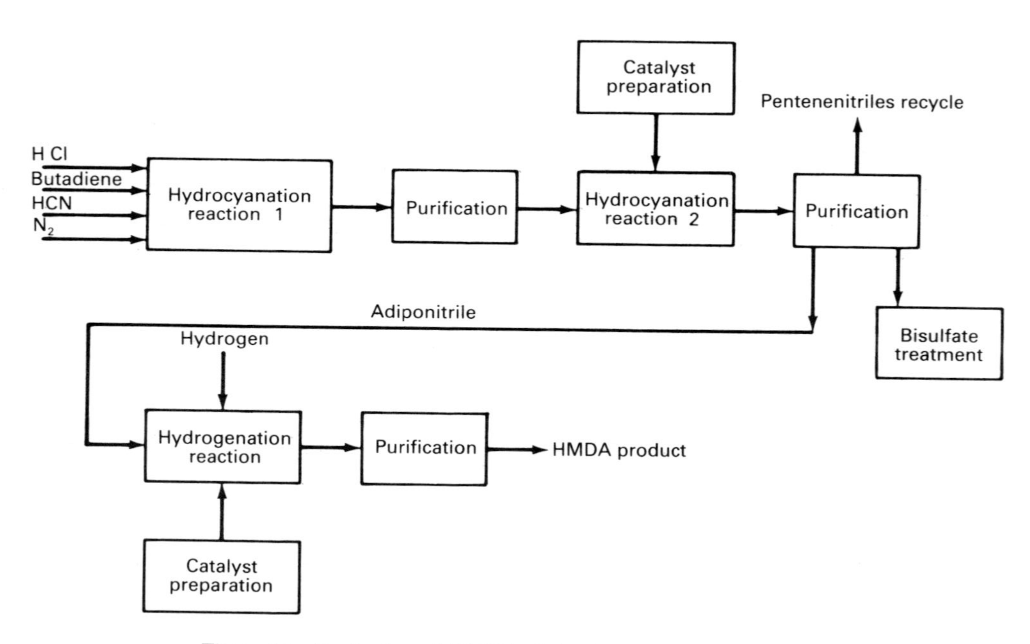

Figure 4.3 Production of HMDA via hydrocyanation of butadiene

anolyte solution contains aqueous sulfuric acid. The following reactions take place:

Cathode: $2CH_2{=}CHCN + 2H_2O + 2e^- \longrightarrow NC(CH_2)_4CN + 2OH^-$

Anode: $H_2O \longrightarrow 2H^+ + \frac{1}{2}O_2 + 2e^-$

Overall: $2CH_2{=}CHCN + H_2O \longrightarrow NC(CH_2)_4CN$

Principal by-products are propionitrile, high-boiling oligomers of acrylonitrile, and bis(cyanoethyl)ether. A block flow diagram for the process is shown in Figure 4-4.

In both the DuPont and Monsanto technologies, capital investments are similar and raw material costs account for about two-thirds of the cost of production. The major raw material for the Monsanto technology is acrylonitrile, whereas the DuPont technology is strongly dependent on costs for both butadiene and hydrogen cyanide. Alternative routes are being studied but are not yet close to commercialization. These routes include a process by ICI that involves catalytically converting acrylonitrile to the linear dimer and the hydrogenation of the dimer to HMDA. The dimerization uses a phosphite catalyst system at temperatures of under 120°C and report yields of over 95 percent to the linear dimer. The potential advantage over Monsanto's technology is claimed to be lower capital investment and utility costs, although yields are somewhat lower. The lower costs are probably due to the use of a catalytic system rather than the more expensive electrolytic system. Halcon is also working on a catalytic process for conversion of acrylonitrile to HMDA. This process is somewhat similar to the ICI technology described previously, except that a branched dimer, 2-methylene-glutaronitrile, is obtained instead of the linear dimer. This stage then requires an isomerization step, which is not required in the ICI technology.

Market

As mentioned previously, HMDA is almost exclusively used in the manufacture of Nylon 6/6, and its fortunes closely follow the fortunes of this nylon. The poor construction market in the early eighties resulted in a drop of over 10 percent from the 1979 peak. The resurging home construction industry in 1983 and 1984 improved the market considerably. Major U.S. producers of HMDA are shown in Table 4-5 (2). Major producers in Western Europe are shown in Table 4-6 (3). Major Japanese producers include Asahi Chemical, Toray, and Hodogaya Chemical.

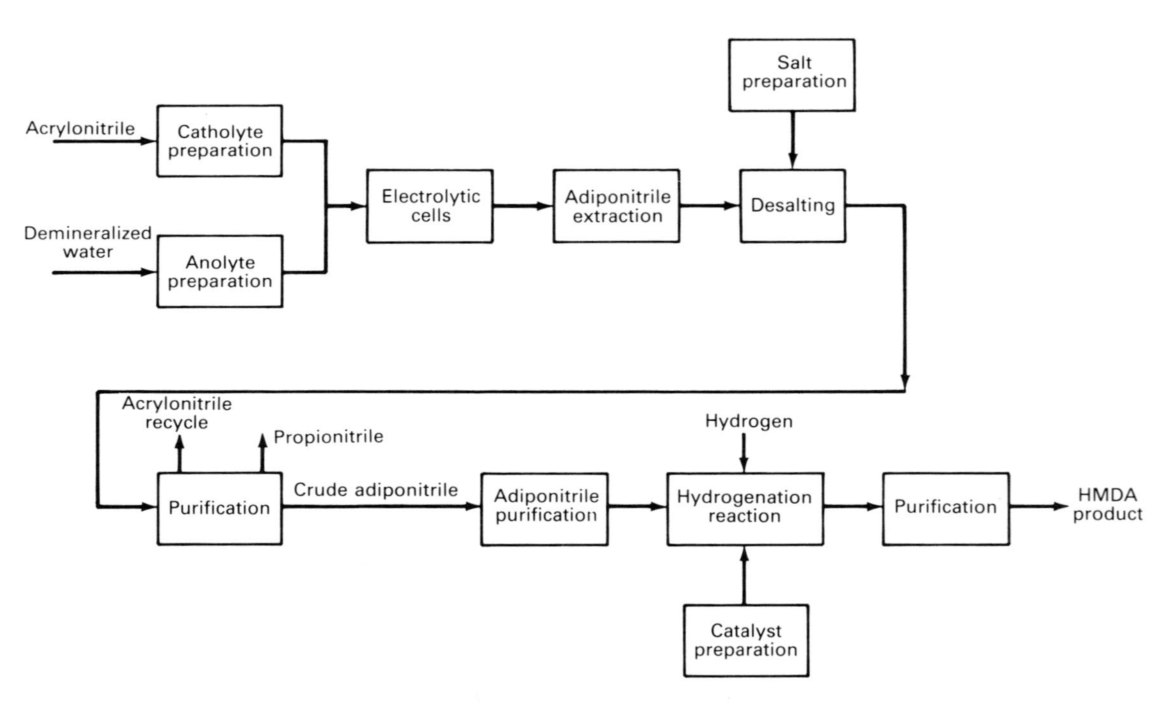

Figure 4.4 Production of HMDA via the electrolytic process

TABLE 4-5 Major U.S. Producers of HMDA

Producer	Location	Estimated Capacity mm lb/yr	m MT/yr
DuPont	Orange, Texas	475	216
	Victoria, Texas	315	143
Monsanto	Decatur, Alabama	200	91
	Pensacola, Florida	200	91
	TOTAL	1190	541

TABLE 4-6 Major Western European Producers of HMDA

Producer	Location	Estimated Capacity mm lb/yr	m MT/yr
BASF	Ludwigshafen, W. Germany	121	55
ICI	Wilton, U.K.	165	75
Polyamide Intermediates	Middlesbrough, U.K.	198	90
Rhone-Poulenc	Chalampe, France	243	110
	Saint-Fons, France	176	80
	TOTAL	903	410

Economics

Typical production costs for hexamethyldiamine from butadiene and from acrylonitrile are as follows:

Via Butadiene

Capacity: 100 mm lb/yr (45 m MT/yr)
Capital cost: BLCC, $43 mm; OSBL, $18 mm; WC, $28 mm

	¢/lb	$/MT	%
Raw materials[a]	50.8	1120	61
Utilities	19.2	423	23
Operating costs	2.8	62	3
Overhead costs	10.4	229	13
Cost of production	83.2	1834	100
Transfer price	101.5	2238	

[a] Butadiene at 32¢/lb; HCN at 35¢/lb

Via Acrylonitrile

Capacity: 100 mm lb/yr (45 m MT/yr)
Capital cost: BLCC, $52 mm; OSBL, $21 mm; WC, $30 mm

	¢/lb	$/MT	%
Raw materials[a]	51.6	1138	58
Utilities	22.7	501	25
Operating costs	2.9	64	3
Overhead costs	12.2	269	14
Cost of production	89.4	1972	100
Transfer price	111.3	2454	

[a] Acrylonitrile at 42¢/lb

REFERENCES

Maleic Anhydride

1. "Chemical Profile," *Chemical Marketing Reporter,* August 1, 1983.
2. List, H. L. "Maleic Anhydride—A Status Report," *International Petrochemical Developments,* March 1, 1981.
3. U.S. Patent 2,625,519 (January 13, 1953) to DuPont.
4. U.S. Patent 2,691,660 (October 12, 1954) to DuPont.
5. U.S. Patent 3,293,268 (December 20, 1966) to Princeton Chemical Research.
6. "A New Maleic Process Awaits Better Times," *Chemical Week,* October 6, 1982, p. 31.
7. *1984 Directory of Chemical Producers—United States,* SRI International.
8. Corbett, J. "Butane Takes Over the Maleic Industry," *Chemical Business,* May 3, 1982, pp. 35–39.
9. Budi, F., A. Neri, and G. Stefani. "Future Maleic Anhydride Keys to Butane," *Hydrocarbon Processing,* January 1982, pp. 159–61.
10. "Petrochemical Handbook '83," *Hydrocarbon Processing,* November 1983.
11. *1984 Directory of Chemical Producers—Western Europe,* SRI International.
12. *Kirk-Othmer Encyclopedia of Chemical Technology,* Third Edition, Vol. 14. New York: Wiley, 1984, pp. 770–93.
13. *JCW Chemicals Guide 82/83.* Japan: The Chemical Daily Co. Ltd., March 1982.
14. "Maleic Makers in Tight Market," *Chemical Marketing Reporter,* April 23, 1984.
15. *Chemical Industry Yearbook,* Second Edition. Surrey, England: Industrial Press, 1984.

Butadiene

1. List, H. L. "Butadiene—A Status Report," *International Petrochemical Developments,* Vol. 3, No. 7, April 1, 1982.
2. *1984 Directory of Chemical Producers—Western Europe,* SRI International.

3. *1984 Directory of Chemical Producers—United States,* SRI International.
4. *Kirk-Othmer Encyclopedia of Chemical Technology,* Third Edition, Vol. 4. New York: Wiley, 1984, pp. 313–37.
5. *JCW Chemicals Guide 82/83.* Japan: The Chemical Daily Co. Ltd., March 1982.
6. *Chemical Industry Yearbook,* Second Edition. Surrey, England: Industrial Press, 1984.
7. "Chemical Profile," *Chemical Marketing Reporter,* April 22, 1985.

Hexamethylenediamine

1. List, H. L. "Hexamethylene Diamine—A Status Report," *International Petrochemical Developments,* Vol. 2, No. 10, May 15, 1981.
2. *1984 Directory of Chemical Producers—United States,* SRI International.
3. *1984 Directory of Chemical Producers—Western Europe,* SRI International.
4. *JCW Chemicals Guide 82/83.* Japan: The Chemical Daily Co. Ltd., March 1982.

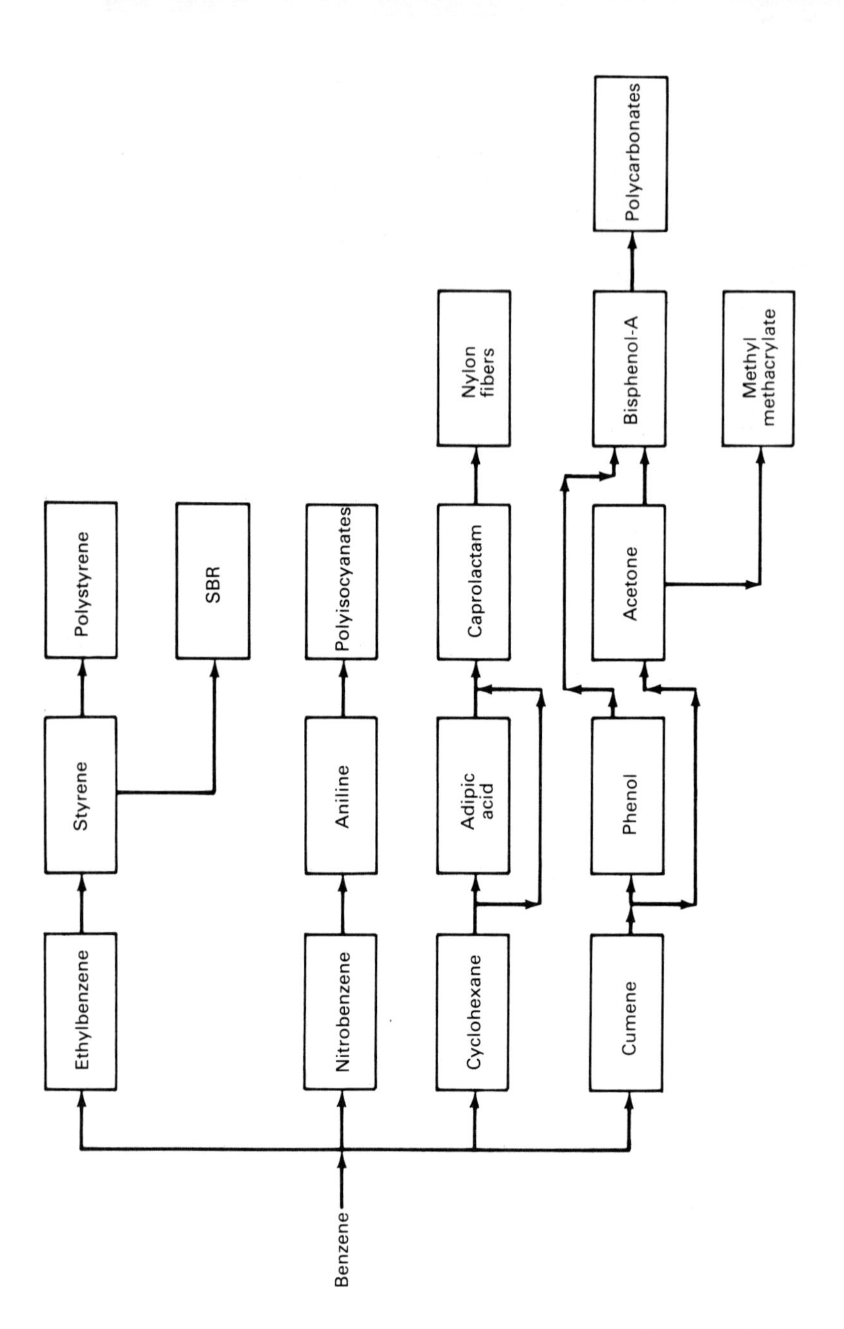

Benzene derivatives

CHAPTER FIVE

Benzene Derivatives

ETHYLBENZENE AND SYTRENE

Production of ethylbenzene and styrene are interrelated, since virtually all of the ethylbenzene production is used to produce styrene monomer. The styrene monomer use covers a broad, diversified range of downstream products.

Technology

A small amount of ethylbenzene is recovered by fractionation of the reformate, but the material is mostly produced from benzene and ethylene via a Friedel-Crafts alkylation as follows:

$$C_6H_6 + CH_2{=}CH_2 \xrightarrow{\text{catalyst}} C_6H_5CH_2CH_3$$

In most cases the catalyst used is aluminum trichloride, with ethylene chloride as a promoter. The reaction is highly exothermic, and the temperature must be controlled, since temperatures above 130°C cause loss of catalyst activity and increased production of polyalkyl aromatics. Conversion to ethylbenzene is essentially dependent on reactor inlet temperature and feedstock composition. The polyalkyl aromatics are easily converted to ethylbenzene by transalkylation with benzene using a crystalline aluminosilicate zeolite as a catalyst. Yields are almost quantitative.

Production of styrene from ethylbenzene involves a straightforward dehydrogenation using an iron oxide catalyst promoted with chromium. The reaction takes place at about 600°C in the presence of superheated steam. Overall yields are generally in excess of 95 percent. The reaction follows:

$$C_6H_5CH_2CH_3 \xrightarrow{H_2O,\ \text{catalyst}} C_6H_5CH{=}CH_2 + H_2$$

The steam serves several functions. It decreases the hydrocarbon partial pressure, thereby favoring the preceding reaction, and it reduces coking of the catalyst. In addition, it provides heat for the endothermic dehydrogenation reaction. A block flow diagram for the process is shown in Figure 5-1.

Another route to styrene is associated with Oxirane's process for the production of propylene oxide. In this case the coproduct styrene is obtained via the dehydration of methyl benzyl alcohol. The rising cost of the feedstock components, primarily benzene, has resulted, and will continue to result, in the shutting down of the older and smaller inefficient plants and the building of world-scale plants using modern technology and resulting in higher yields. Many of these units will use the Mobil-developed ethylbenzene process, employing a highly selective zeolite catalyst. The rising cost of benzene has also resulted in

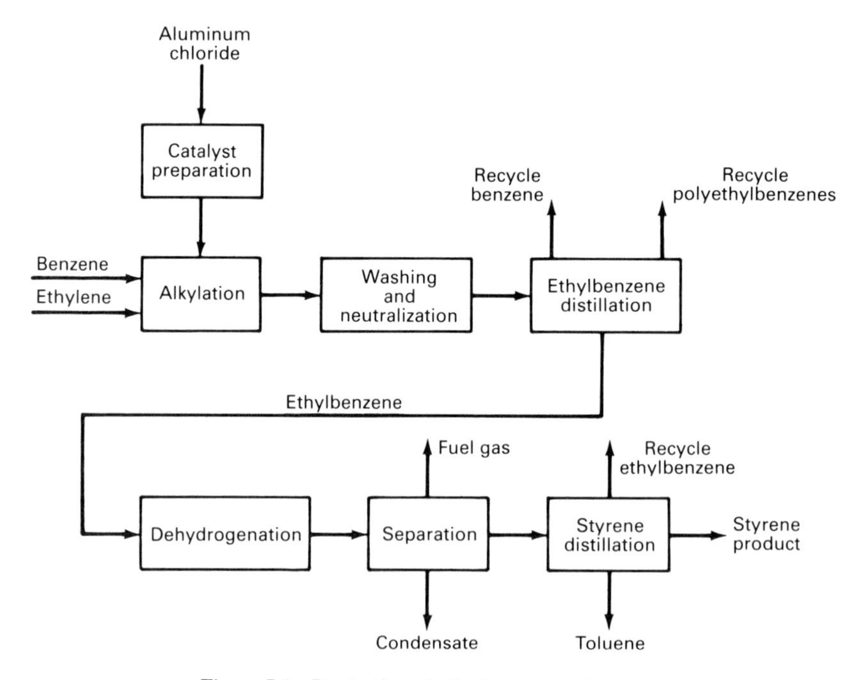

Figure 5.1 Production of ethylbenzene and styrene

research efforts aimed at starting with toluene in place of benzene. Monsanto has been active in this area and has concentrated on the following two approaches:

1. Production of styrene from toluene in a two-step process involving dimerization of toluene to stilbene, followed by disproportionation with ethylene. The dimerization uses a lead oxide catalyst system.
2. Production of styrene by the alkylation of toluene with methyl alcohol (12). A zeolite catalyst incorporating some cesium or rubidium is used with a boron promoter. This route could eventually offer a process with coal, via synthesis gas, as feedstock.

Market

The market for styrene monomer (and therefore ethylbenzene) has the advantage of serving a fairly diverse group of downstream products, some of which are expected to grow fairly rapidly. The approximate portions of styrene for various end uses in the United States are shown in the following table (2,3):

End Uses	%
Polystyrene	52
ABS (acrylonitrile-butadiene-styrene)	9
SBR (styrene-butadiene rubber)	7
SB (styrene-butadiene latex)	6
Unsaturated polyester resins	6
Export	15
Other	5

Polystyrene and unsaturated polyester resins are expected to grow faster than all other end use markets. U.S. demand for styrene in 1983 was about 6.4 billion pounds and is expected to rise to almost 8 billion pounds by 1987. As can be seen from the preceding table, polystyrene resins, which include rubber-modified impact grades, general purpose grades, and foam, represent the largest end use for styrene. The impact grades are used in applications such as home appliances and drain pipes. General purpose grades are used in products such as toys and light diffusers. Polystyrene foam is used in disposable drinking cups and trays as well as in packaging applications. This market, as well as that of unsaturated polyester resins, is generally expected to grow at an average annual rate of 8 percent over the next few years, with ABS and SBR following closely behind.

Much of the unsaturated polyester resins are used in glass-fiber-reinforced plastics, which are being increasingly substituted for metal components to meet the U.S. federal government's automobile fuel economy standards. Since the end uses of styrene discussed previously account for about three-fourths of the demand, the next few years appear promising and the tightening supply is expected to result in additional facilities, probably constructed by the current manufacturers.

The export percentage shown in the preceding table is expected to decrease continuously for two reasons: (1) The raw material advantages in the United States are ending and (2) additional styrene capacity is planned overseas, including Saudi Arabia and Western Canada. Current producers of both ethylbenzene and styrene in the United States are shown in Table 5-1 (4).

USS has a 125 million pound per year ethylbenzene facility and a 120 million pound per year styrene facility at Houston, Texas, which are both idle. Hoechst has a 600 million pound per year ethylbenzene plant at Baton Rouge, Louisiana, which is idle. Koch has a 60 million pound per year styrene facility at Corpus Christi, Texas, which is idle. Cos-Mar, owned by Cosden and Borg-Warner, and Gulf are expected to expand capacity.

Major producers of ethylbenzene and styrene in Western Europe are shown in Table 5-2 (13). BASF is expected to shut down its ethylbenzene facility in Belgium, and CdF Chimie is expected to expand its ethylbenzene capacity at Carling, France, by about 210 million pounds per year (95,000 MT/yr). In addition, Companhia Nacional de Petroquimica is building a facility at Sines, Portugal, for ethylbenzene. Start-up of the 386 million pound per year (175,000 MT/yr) is scheduled for 1989.

Major producers of styrene in Japan are shown in Table 5-3.

A recent study has indicated that most of the growth in capacity for styrene will be in the developing countries, which will significantly change the worldwide distribution of styrene capacity. The current and projected worldwide distributions of styrene capacity and styrene demand are shown in Tables 5-4 and 5-5, respectively (6). Note that demand is not expected to change as significantly as capacity.

TABLE 5-1 Current U.S. Producers of Ethylbenzene and Styrene

Producer	Location	Estimated Capacity, mm lb/yr (m MT/yr)			
		Ethylbenzene		Styrene	
American Hoechst	Bayport, Texas	1050	(476)	900	(408)
Amoco	Texas City, Texas	900	(408)	790	(358)
Arco	Channelview, Texas	1200	(544)	1000	(454)
	Beaver Valley, Pennsylvania	—		220	(100)
Cos-Mar	Carville, Louisiana	1730	(785)	1300	(590)
Dow	Freeport, Texas	1850	(839)	1400	(635)
	Midland, Michigan	—		300	(136)
DuPont	Chocolate Bayou, Texas	60	(27)	—	
El Paso	Odessa, Texas	280	(127)	250	(113)
Gulf	St. James, Louisiana	680	(308)	600	(272)
Koch	Corpus Christi, Texas	100	(45)	—	
Monsanto	Texas City, Texas	1600	(726)	1500	(680)
	TOTAL	9450	(4285)	8260	(3746)

TABLE 5-2 Major Western European Producers of Ethylbenzene and Styrene

Producer	Location	Estimated Capacity, mm lb/yr (m MT/yr)			
		Ethylbenzene		Styrene	
BASF	Antwerp, Belgium	882	(400)	—	
ATOCHEM	Gonfreville-L'Orcher, France	595	(270)	551	(250)
CdF Chimie	Carling, France	584	(265)	662	(300)
Elf-Aquitaine	Laco, France	143	(65)	110	(50)
BASF	Ludwigshafen, W. Germany	662	(300)	992	(450)
Chemische Werke Huels	Marl, W. Germany	573	(260)	485	(220)
Rheinische Olefinwerke	Wesseling, W. Germany	948	(430)	926	(420)
Deutsch Texaco	Heide, W. Germany	24	(11)	—	
Montedison	Priolo, Italy	154	(70)	—	
Sares Chimica	Sarroch, Italy	88	(40)	—	
Anic	Ravenna, Italy	—		99	(45)
Dow Chemical	Terneuzen, Netherlands	2205	(1000)	1544	(700)
Shell Chemie	Moerdijk, Netherlands	827	(375)	717	(325)
Montoro-Empresa	Puertollano, Spain	265	(120)	220	(100)
Forth Chemicals	Port-Talbot, Wales, U.K.	573	(260)	485	(220)
ICI	Middlesbrough, U.K.	99	(45)	—	
	TOTAL	8622	(3911)	6791	(3080)

TABLE 5-3 Major Japanese Producers of Styrene

Producer	Location	Estimated Capacity	
		mm lb/yr	m MT/yr
Mitsubishi Petrochemical	Yokkaichi, Kashima	900	408
Asahi Chemical	Mizushima, Kawasaki	870	395
Sumitomo Chemical	Chiba	710	322
Denki	Chiba	350	159
Mitsui Toatsu	Sakai	200	91
Nippon Steel Chemical	Tobata, Ohita	370	168
Shin-Daikyowa	Yokkaichi	200	91
Idemitsu	Chiba	330	150
	TOTAL	3930	1784

TABLE 5-4 Distribution of World Styrene Capacity

	1981	1985
North America, Western Europe, Japan	84	75–79
Eastern Europe	10	12–12.5
Other	6	8.5–13

TABLE 5-5 Distribution of World Styrene Demand

	1982	1985
North America, Western Europe, Japan	79	77
Eastern Europe	11	11.5
Other	10	11.5

A possible styrene replacement being developed is para-methylstyrene (8). Mobil Chemical has built a 35 million pound per year facility and is producing limited quantities of the polymer poly-para-methylstyrene. They have projected that by 1990 this material could have 10 to 20 percent of the styrene market in the United States. The material can be made from toluene and ethylene rather than from benzene and ethylene as in the case of styrene.

Economics

Typical production costs for ethybenzene and styrene are as follows:

Ethylbenzene from Benzene

Capacity: 1200 mm lb/yr (544 m MT/yr)
Capital cost: BLCC, $34 mm; OSBL, $17; WC, $84

	¢/lb	$/MT	%
Raw materials[a]	18.6	410	89
Utilities	0.6	13	3
Operating costs	0.3	7	1
Overhead costs	1.5	33	7
Cost of production	21.0	463	100
By-product (steam)	0.8	18	
Net cost of production	20.2	445	
Transfer price	21.5	474	

[a] Ethylene at 22¢/lb; benzene at 17¢/lb

Styrene from Ethylbenzene

Capacity: 750 mm lb/yr (340 m MT/yr)
Capital cost: BLCC, $59 mm; OSBL, $35 mm; WC, $74 mm

	¢/lb	$/MT	%
Raw materials[a]	24.7	545	81
Utilities	2.3	51	8
Operating costs	0.5	11	2
Overhead costs	2.7	60	9

Styrene from Ethylbenzene (*continued*)

	¢/lb	$/MT	%
Cost of production	30.2	667	100
By-product[b]	0.8	18	
Net cost of production	29.4	649	
Transfer price	33.2	732	

[a] Ethylene at 22¢/lb

[b] Toluene at 17¢/lb

POLYSTYRENE

Polystyrene is the major derivative of styrene. High-impact, general purpose, and expandable polystyrenes account for almost two-thirds of styrene consumption. Other major derivatives of styrene are synthetic rubber, copolymer resins, and unsaturated polyesters.

Technology

General purpose and high-impact polystyrene are the largest end uses, accounting for well over half the polystyrene demand in the United States, Western Europe, and Japan. Most of this demand is for high-impact grades, which are modified by incorporating rubber into the polystyrene.

Successful commercial production of polystyrene was started in the United States by Dow Chemical in 1937. Several variations in the polymerization technology are offered, but the processes are mostly quite similar. Most of the processes use mass polymerization or suspension polymerization, either batch or continuous. The processes vary primarily in the design of the prepolymerization and polymerization reactors. The reaction is controlled at a specific temperature and degree of polymerization to yield a product with the desired physical properties. The reaction rate is thermally controlled. After polymerization the volatiles are removed, usually under vacuum. Additives are incorporated to impart specific properties.

In rubber-modified polystyrenes the rubber is dispersed in the polystyrene matrix in the form of discrete particles. The rubber toughens the otherwise brittle polystyrene. The use of butadiene rubbers is particularly effective when the rubber is present during the polymerization of styrene. Grafting of some of the styrene to the rubber takes place. Some of the optional additives include catalysts, chain transfer agents, and mineral oil, all designed to produce the desired balance between production economics and end product properties. The usual range of rubber content is up to about 15 percent. A block flow diagram for a typical process is shown in Figure 5-2.

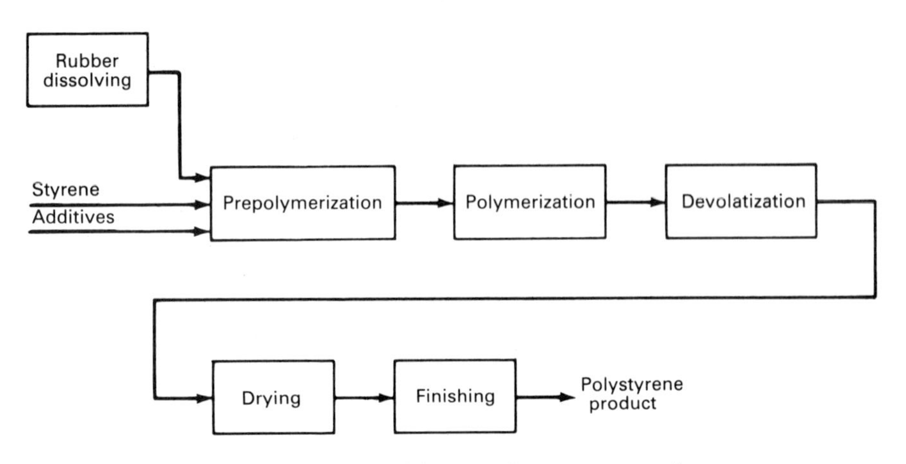

Figure 5.2 Production of high-impact polystyrene—suspension process

Market

Polystyrene has shown continuous growth but has slowed down since the mid-seventies. Growth is sluggish for several reasons, including slow economic growth and competition from other polymers, such as polypropylene. No additional major end uses for polystyrene appear to be developing. The major market for polystyrene is in the packaging area. This application probably accounts for close to half of the consumption, including disposable cups, clear meat trays, plastic cutlery, and coffee lids. The packaging area is where polypropylene has come on strong as a substitute.

Another major application is in appliances and electrical products, including radio and television cabinets and back covers, parts of refrigerators, and housings for small appliances. Although ABS and polypropylene are competitive, polystyrene is generally expected to hold its present market.

The third largest end use of polystyrene is for toys and housewares. Since polystyrene components can be more easily assembled than polypropylene components, polystyrene is expected to hold a good part of the present market.

Expandable polystyrene has had stronger growth than other forms of polystyrene and now accounts for close to 15 percent of styrene demand in the United States, Western Europe, and Japan. Almost half of this demand is in the construction area, including insulation. Use of this material has also grown dramatically in food packaging.

U.S. demand for polystyrene in 1983 was approximately 3.3 billion pounds. This amount was down from a level of about 3.6 billion pounds in both 1980 and 1981. Authorities project that demand could grow to perhaps 6 billion pounds by the end of the decade. Exports of polystyrene are relatively small, about 100 million pounds in 1982.

Table 5-6 lists the major U.S. producers of polystyrene and their locations and estimated capacities (3,6).

Major Western European producers of polystyrene resins, including impact-modified and expandable resins, are shown in Table 5-7 (5). Major Japanese producers are shown in Table 5-8.

TABLE 5-6 Major U.S. Producers of Polystyrene

Producer	Location	Estimated Capacity	
		mm lb/yr	m MT/yr
American Hoechst	Chesapeake, Maryland	300	136
	Peru, Illinois	380	172
Arco	Beaver Valley, Pennsylvania	550	249
BASF Wyandotte	Jamesburg, New Jersey	220	100
Cosden	Big Spring, Texas	45	20
	Calumet City, Illinois	270	122
	Windsor, New Jersey	120	54
Dow	Gales Ferry, Connecticut	190	86
	Ironton, Ohio	190	86
	Joliet, Illinois	140	63
	Midland, Michigan	220	100
	Torrance, California	200	91
Georgia Pacific	Painesville, Ohio	71	32
Gulf Oil	Marietta, Ohio	310	141
Huntsman	Belpre, Ohio	300	136
Mobil Chemical	Holyoke, Massachusetts	90	41
	Joliet, Illinois	300	136
	Santa Ana, California	65	29
Monsanto	Addyston, Ohio	300	136
	Decatur, Alabama	100	45
	Springfield, Massachusetts	300	136
Polysar	Akron, Ohio	120	54
	Leominster, Massachusetts	180	82
Amoco Chemicals	Joliet, Illinois	300	136
	Torrance, California	35	16
	Willow Springs, Illinois	90	41
Texstyrene Plastics	Fort Worth, Texas	100	45
	TOTAL	5486	2485

TABLE 5-7 Major Western European Producers of Polystyrene

Producer	Location	Estimated Capacity	
		mm lb/yr	m MT/yr
BASF	Antwerp, Belgium	364	165
Dow Chemical	Tessenderlo, Belgium	132	60
Montefina	Feluy, Belgium	132	60

TABLE 5-7 Major Western European Producers of Polystyrene (*continued*)

Producer	Location	Estimated Capacity	
		mm lb/yr	m MT/yr
Neste Oy	Kulloo, Finland	44	20
Duomrn PS Tehdas Oy	Kokemaki, Finland	24	11
ATOCHEM	Gonfreville-L'Orcher, France	331	150
CdF Chimie	Dieuze, France	243	110
	Ribecourt-Dreslincourt, France	260	118
Dispersions Plast.	Villers-Saint-Paul, France	99	45
Produits Chim. BP	Wingles, France	331	150
Shell Chimie	Berre-L'etang, France	110	50
BASF	Ludwigshafen, W. Germany	1147	520
Chemische Werke Huels	Marl, W. Germany	430	195
Dow Chemical	Lavrion, Greece	66	30
Dow Chemical	Livorno, Italy	110	50
Enichem Polimeri	Porto-Torres, Italy	154	70
Massucchelli Cell.	Castiglione-Olona, Italy	33	15
Montepolimeri	Mantova, Italy	595	270
Brabant	Etten-Leur, Netherlands	22	10
Dow Chemical	Terneuzen, Netherlands	198	90
Hoechst	Breda, Netherlands	243	110
Brodr. Sunde	Spjelkavik, Norway	20	9
Aiscondel	Monzon-del-Rio-Cinca, Spain	29	13
Arrahona	Prat-de-Llobregat, Spain	154	70
BASF	Tarragona, Spain	77	35
Dow Chemical	Axpe-Erandio, Spain	99	45
Svenska Polystyren	Trelleborg, Sweden	66	30
ATO	Stalybridge, U.K.	176	80
Dow Chemical	Barry, Wales, U.K.	176	80
Shell Chemicals	Carrington, U.K.	159	72
	TOTAL	6024	2733

TABLE 5-8 Major Japanese Producers of Polystyrene

Producer	Location	Estimated Capacity	
		mm lb/yr	m MT/yr
Asahi Chemicals	Kawasaki, Mizushima, Chiba	555	252
Mitsubishi-Monsanto	Yokkaichi	280	127
Denka	Chiba	205	93
Nippon Polystyrene	Kawasaki	130	59
Idemitsu	Chiba	200	91
Dainippon Ink Chemical	Chiba	130	59
Toyo Polystyrene	Kawasaki, Sakai	260	118
	TOTAL	1760	799

Economics

Production costs for a typical high-impact polystyrene process via bulk suspension are as follows:

Capacity: 220 mm lb/yr (100 m MT/yr)
Capital cost: BLCC, $36 mm; OSBL, $15 mm; WC, $34 mm

	¢/lb	$/MT	%
Raw materials[a]	37.5	827	80
Utilities	2.2	49	5
Operating costs	1.5	33	3
Overhead costs	5.4	119	12
Cost of production	46.6	1028	100
Transfer price	53.6	1182	

[a] Styrene at 35¢/lb

ABS

Acrylonitrile-butadiene-styrene resins (ABS) have had good growth and acceptability as a versatile resin. Many common ratios are used to obtain a substantial range of physical properties. For most products, processing is by injection molding or extrusion. Although ABS is normally opaque, clear grades can be produced. It is also possible to produce grades that can be plated with metal. ABS is also compatible with PVC and polycarbonates, and some specialty blends are available.

Technology

Most of the worldwide production of ABS is accomplished by emulsion polymerization and bulk suspension processes.

Emulsion polymerization. This process basically involves the free radical polymerization of a mixture of polybutadiene latex with styrene-acrylonitrile. The process can be either batch or continuous, but there is substantial evidence that batch operation is gradually being eliminated. The styrene and acrylonitrile are in water with an initiator, an emulsifier, and a chain length regulator. The emulsion is heated and agitated to accomplish the graft polymerization. The water is the continuous phase and the reactants the discontinuous phase. As the copolymerization proceeds, the resin particles that are formed are suspended in the water phase.

Bulk suspension process. This technology involves a two-stage copolymerization process. In the first stage polybutadiene or another elastomer is dissolved in a mixture of styrene and acrylonitrile and is bulk prepolymerized. After conversion of about 40 percent, the viscous mass is dispersed in water and the copolymerization is completed. Monomer-soluble, free radical initiators are used, and chain length regulators are also added. Initially the elastomers in styrene and acrylonitrile are in one continuous phase, but as the reaction proceeds, the styrene-acrylonitrile copolymer forms a dispersed phase. Vigorous agitation transfers the elastomer to the dispersed phase. The product from the bulk stage of the process is dispersed in water and kept in suspension with the aid of a protective colloid, such as polyvinyl alcohol, and the copolymerization is completed in the presence of a free radical initiator. Considerable work, most of it kept highly confidential, has been directed toward development of a continuous suspension process. The major problem is the agglomeration of the sticky beads. Therefore agitation is a prime concern in the design of a continuous suspension process. A typical block flow diagram is shown in Figure 5-3.

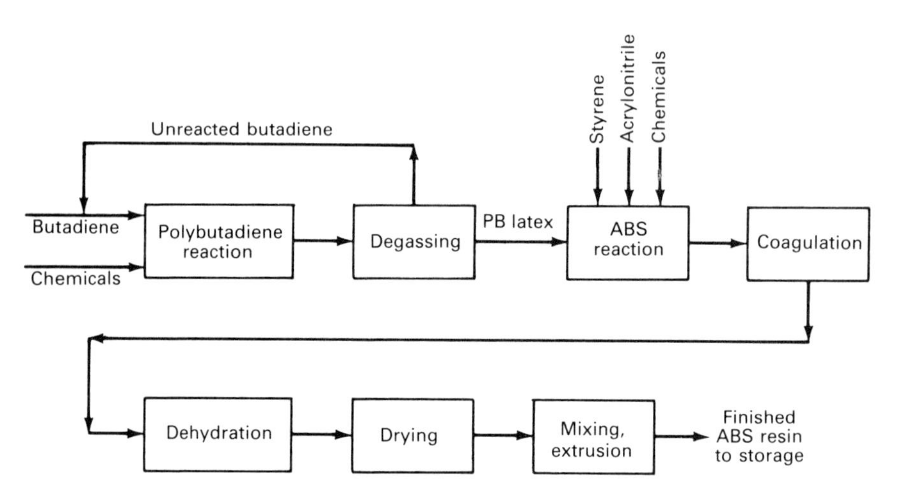

Figure 5.3 Production of ABS resins

Market

In the United States the construction, transportation, and electrical markets account for almost three-fourths of the total consumption of ABS (3). In the early eighties the market was hard hit by the worldwide recession in the major markets—automobiles and appliances. Since 1979, several producers, including USS Chemicals, dropped out of the market. Production dropped from about 1.1 billion pounds in 1979 to about 800 million pounds in 1982. In addition to the recession, the ABS pipe market has declined markedly because of displacement by PVC (4). The major reason for the displacement has been the price increases for ABS, which have exceeded those for other plastics. At present three major producers are in the United States. Locations and estimated capacities are shown in Table 5-9 (2,9). All the major producers have secure raw material positions. Monsanto produces acrylonitrile, Dow and Monsanto produce their own styrene, and Borg-Warner owns half of a joint venture that produces styrene.

In Western Europe the major end uses of ABS are in appliances and automobiles, accounting for close to half of the total consumption. Over three-fourths of the total demand for these products is in four countries: West Germany, France, Italy, and the United Kingdom. Major producers in Western Europe are shown in Table 5-10 (7). Major uses in Japan are in the automotive, electrical, and mechanical applications. Major Japanese producers are shown in Table 5-11.

TABLE 5-9 Major U.S. Producers of ABS

Producer	Location	Estimated Capacity	
		mm lb/yr	m MT/yr
Borg-Warner	Ottowa, Illinois	235	107
	Port Bienville, Mississippi	150	68
	Washington, W. Virginia	335	152
Dow	Allyn's Point, Connecticut	70	32
	Hanging Rock, Ohio	70	32
	Midland, Michigan	80	36
	Torrance, California	80	36
Monsanto	Addyston, Ohio	355	161
	Muscatine, Iowa	115	52
	Springfield, Massachusetts	30	14
	TOTAL	1520	690

TABLE 5-10 Major Western European Producers of ABS

Producer	Location	Estimated Capacity	
		mm lb/yr	m MT/yr
Monsanto	Antwerp, Belgium	154[a]	70
CdF Chimie	Villers-Saint-Sepulcre, France	110	50
BASF	Ludwigshafen, W. Germany	132	60
Bayer	Dormagen, W. Germany	165	75
	Tarragona, Spain	44	20
Anic	Porto-Torres, Italy	29[a]	13
	Ravenna, Italy	99	45
Enichimica	Ferrara, Italy	79	36
Borg-Warner	Amsterdam, Netherlands	154	70
	Grangemouth, U.K.	154	70

TABLE 5-10 Major Western European Producers of ABS (*continued*)

Producer	Location	Estimated Capacity mm lb/yr	m MT/yr
Dow Chemical	Terneuzen, Netherlands	88	40
DSM	Geleen, Netherlands	110	50
Aiscondel	Monzon-del Rio Cinca, Spain	33	15
ATO Chemical	Stalybridge, U.K.	11	5
Enoxy Chemical	Hythe Southampton, U.K.	22	10
	TOTAL	1384	629

[a] Includes SAN

TABLE 5-11 Major Japanese Producers of ABS

Producer	Location	Estimated Capacity mm lb/yr	m MT/yr
Asahi	Mazahima	55	25
Denka	Goi	145	66
Kanegafuchi	Takasazo	80	36
Mitsubishi-Monsanto	Yokkachi	155	70
Mitsubishi Rayon	Ohtake	170	77
Nippon Zeon	Nichama	66	30
Sumitomo	Nichama	95	43
Toray	Chiba	120	54
Ube Cycon	Ube	130	59
	TOTAL	1015	460

Economics

Production costs for a typical continuous emulsion polymerization process are as follows:

Capacity: 150 mm lb/yr (68 m MT/yr)
Capital cost: BLCC, $32 mm; OSBL, $13 mm; WC, $28 mm

	¢/lb	$/MT	%
Raw materials[a]	41.6	917	75
Utilities	3.1	68	6
Operating costs	2.9	64	5
Overhead costs	7.8	172	14
Cost of production	55.4	1221	100
Transfer price	64.4	1420	

[a] Styrene at 35¢/lb; butadiene at 32¢/lb; acrylonitrile at 45¢/lb

SBR

Styrene-butadiene rubber (SBR) accounts for the majority of synthetic rubber production, although its share of the total market is decreasing. It is the most important synthetic rubber used by the tire industry because of its good mechanical and physical properties and its favorable cost.

Technology

SBR is manufactured by copolymerizing about 75 percent butadiene and 25 percent styrene. The polymerization is accomplished either in an emulsion using free radical initiators, such as peroxides, or in a solution using anionic initiators or coordination catalysts. When polymerization takes place in an emulsion, a random copolymer is produced. Polymerization of butadiene and styrene in solution usually produces block copolymers when both monomers are charged with the anionic or coordination catalyst. Butadiene polymerizes first; when it is virtually depleted, styrene begins to polymerize. Figure 5-4 shows a block flowsheet for a typical emulsion polymerization process.

The development of more effective catalysts and activators has permitted the polymerization reaction to be carried out at lower temperatures, even somewhat below ambient. The material produced at lower temperatures is generally considered to have superior properties, and most of the emulsion production currently in commercial use operates at the lower temperatures. Since the polymerization is exothermic, refrigeration is generally required to maintain the temperature. The conversion is usually limited to about 60 percent to prevent formation of gel and is generally reached in about 12 hours. The butadiene and

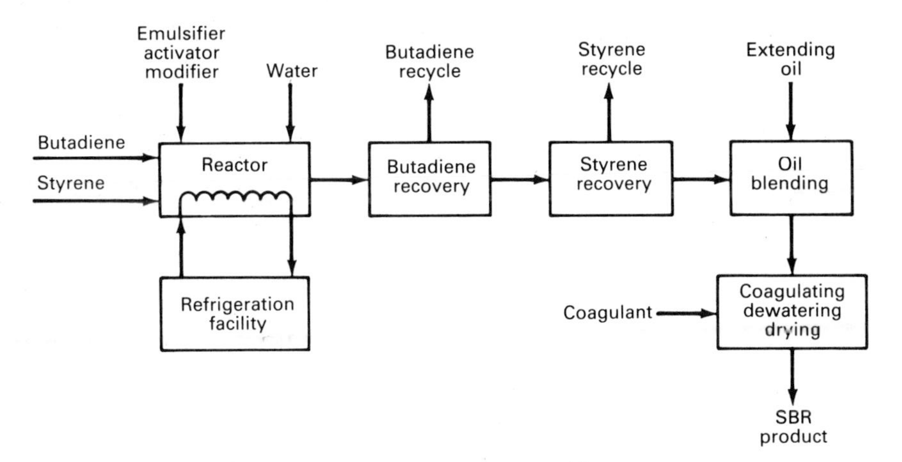

Figure 5.4 Production of SBR—emulsion process

styrene are removed from the latex by vacuum and steam stripping and recycled after purification. If oil-extended grades are being produced, emulsified oil is blended into the stripped latex. The solid product is recovered by coagulating the latex and then dewatering and drying.

Market

Total U.S. consumption of synthetic rubber in 1982 was in excess of 1.6 million metric tons, of which about 56 percent was SBR (4). In 1983 the figure was over 1.8 million metric tons (6). This amount compares with a figure of over 2.3 million metric tons in 1979. Through the end of this decade, growth in annual consumption is estimated at 3.3 percent (5). Global consumption excluding the communist countries is approximately 5 million metric tons and is projected to reach 5.6 million metric tons by 1988 according to the projection by the International Institute of Synthetic Rubber Producers. Regarding SBR specifically, worldwide production ranged from almost 3 million metric tons in 1979 to almost 2.3 million metric tons in 1982. Total synthetic rubber capacity in 1984 was about 8.7 million metric tons, with over a third of this amount in North America and a third in Western Europe. SBR capacity accounted for about 60 percent of the total.

The results of the oil price increases—inflation, smaller cars with smaller tires, a decrease in driving, and a decline in auto production—all combined to reduce rubber demand. In addition, the wider acceptance of radial tires, which use less SBR, added to the difficulty. Radial tires use more natural rubber than synthetic rubber. The nontire uses of SBR will probably grow at a higher rate than the tire uses, but there is substantial competition from other elastomers. The past decade has seen a number of shutdowns, which have decreased SBR capacity, and this trend is expected to continue for the near term. Since a substantial overcapacity exists, new facilities are unlikely to be built in the United States or Western Europe for a considerable period of time.

Table 5-12 shows major producers of styrene-butadiene rubber in the United States (1). In addition, American Synthetic Rubber has a 55,000 metric ton per year mothballed facility at Louisville, Kentucky, using the emulsion polymerization process. In addition to the rubber itself, several companies produce SBR latex, with a total capacity of about 70,000 metric tons per year. These companies include several of these listed in Table 5-12 as well as W. R. Grace, Occidental Petroleum, and Polysar.

Table 5-13 lists major Western European producers of styrene-butadiene rubber, including latex product (2). Major Japanese producers of dry rubber include Japan Synthetic Rubber, Mitsubishi Chemical Industries, Nippon Zeon, and Sumitomo Chemical. In addition, latex is produced by Asahi-Dow, Mitsui Toatsu Chemicals, Dainippon Ink and Chemicals, and Takeda Chemical Industries, as well as by the producers of the dry rubber.

TABLE 5-12 Major U.S. Producers of Styrene-Butadiene Rubber

Producer	Location	Estimated Capacity mm lb/yr	Estimated Capacity m MT/yr	Process
Copolymer Rubber	Baton Rouge, Louisiana	276	125	Emulsion
Firestone	Lake Charles, Louisiana	265	120	Solution
General Tire	Odessa, Texas	198	90	Emulsion
B. F. Goodrich	Port Neches, Texas	295	134	Emulsion
Goodyear Tire	Houston, Texas	714	324	Emulsion
Phillips Petroleum[a]	Borger, Texas	198	90	Emulsion
Synpol	Port Neches, Texas	364	164	Emulsion
	TOTAL	2310	1047	

[a]Phillips shut down its facility in 1984.

TABLE 5-13 Major Western European Producers of Styrene-Butadiene Rubber

Producer	Location	Estimated Capacity mm lb/yr	Estimated Capacity m MT/yr	Product
Donau Chemie	Zwentendorf, Austria	22	10	Latex
Petrofina	Antwerp, Belgium	88	40	Solid
Buna France	Lillebonne, France	51	23	
	Port Jerome, France	26	12	Latex
Polysar France	LaWantzenau, France	232	105	Both
Shell Chimie	Berre-L'Etang, France	187	85	Solid
BASF	Ludwigshafen, W. Germany	33	15	Latex
Bayer	Leverkusen, W. Germany	55	25	Latex
Bunawerke Huls	Marl, W. Germany	397	180	Both
Chemische Werke Huls	Marl, W. Germany	218	99	Both
Enichem Polimeri	Porto-Torres, Italy	176	80	Solid
	Ravenna, Italy	474	215	Both
Polysar Nederland	Arnhem, Netherlands	15	7	Latex
Shell	Rotterdam, Netherlands	298	135	Solid
Calatrava	Santander, Spain	187	85	Solid
Bayer	Bromsgrove, U.K.	11	5	Latex
Doverstrand	Stallingborough, U.K.	40	18	Latex
Enoxy Chemical	Grangemouth, U.K.	88	40	Solid
	Hythe Southampton, U.K.	496	225	Both
	TOTAL	3094	1404	

Economics

Typical production costs for a low-temperature emulsion facility for SBR are as follows:

Capacity: 220 mm lb/yr (100 m MT/yr)
Capital cost: BLCC, $65 mm; OSBL, $26 mm; WC, $30 mm

	¢/lb	$/MT	%
Raw materials[a]	28.3	624	70
Utilities	3.8	84	9
Operating costs	1.7	37	4
Overhead costs	6.6	146	17
Cost of production	40.4	891	100
Transfer price	52.8	1164	

[a] Butadiene at 32¢/lb; styrene at 35¢/lb

NITROBENZENE

Nitrobenzene is one of the more important raw materials for the dye industry, and much of it is used directly or as aniline in dye manufacture. A large amount of aniline is also used in the manufacture of isocyanates. Nitrobenzene is manufactured on a large scale only by aniline manufacturers.

Technology

Nitrobenzene was first manufactured commercially in England in 1856 and therefore has been a long-available industrial chemical. The usual method for the manufacture of nitrobenzene is by the direct nitration of benzene using a nitric acid–sulfuric acid mixture. The reaction vessels are usually cast iron or steel kettles supplied with efficient agitation. The reaction is strongly exothermic, and great care must be taken to provide sufficient cooling for proper temperature control.

The reaction, usually in yields of better than 95 percent, is as follows:

$$C_6H_6 + HNO_3 \xrightarrow{H_2SO_4} C_6H_5NO_2 + H_2O$$

Although the process can be accomplished batchwise, the more-advanced processes for license, for example, by Biazzi, Meissner, and Canadian Industries, are continuous. The processes are quite similar, the major differences being the nitration reactor design and the number of reactors.

After the reaction the spent acid is settled from the crude nitrobenzene and recovered and recycled. The crude nitrobenzene can be used directly in the manufacture of aniline or further purified if it is to be an end product. Possible improvements in the processing have generally moved in two directions. One process involves the use of tubular reactors. The efficiency of mixing, flow properties in the tubular reactor, heat exchange, reaction temperature, and residence time are all important in the potential for this process. Some work has been done using only nitric acid rather than a nitric-sulfuric mixture. A block flow diagram of a typical process is shown in Figure 5-5.

The second improvement involves an "adiabatic" process, which has been jointly developed by American Cyanamid and Canadian Industries and is offered commercially. In this process 65 percent rather than 98 percent is used. This concentration permits the solution to adiabatically absorb the exothermic heat of reaction and eliminates the need for cooling coils in the reactor. In addition, by permitting premixing of the entire charge of benzene, the process reduces the undesirable secondary conversion of mononitrobenzene to dinitrobenzene. Furthermore the aqueous sulfuric acid can be recycled without reconcentration, thereby simplifying the acid-handling operation.

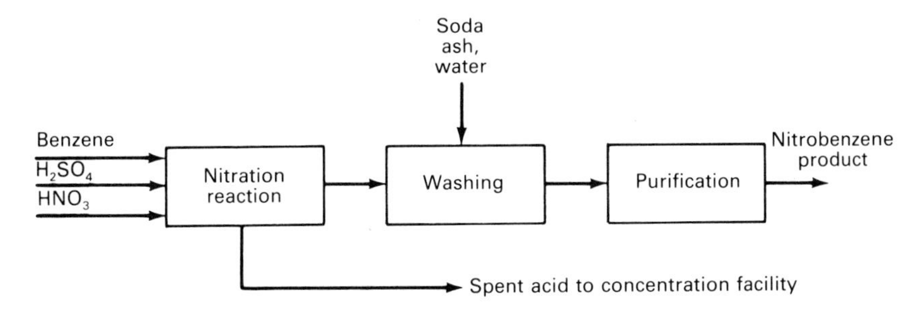

Figure 5.5 Production of nitrobenzene

Market

The largest outlet for nitrobenzene is in the production of aniline, which is the feed for the manufacture of polymeric isocyanates. Arco Chemical uses a different technology for polymeric isocyanates based on a reaction of nitrobenzene with carbon monoxide. Other than Arco all polymeric isocyanates in the United States are made from aniline. Total U.S. demand for nitrobenzene in 1983 was about 910 million pounds, with the expectation that by the mid-eighties it will reach close to a billion pounds. U.S. producers and estimated capacities are as indicated in Table 5-14 (2).

All plants are integrated with adjacent aniline facilities. Arco has repeatedly delayed its polymeric isocyanates facility and lowered its capacity. Since its process is based on the reaction of nitrobenzene and carbon monoxide, the required nitrobenzene facilities, originally scheduled for a capacity of about 260 million pounds per year and 1982 start-up have also been delayed. Mobay is operating at about 55 million pounds per year to serve its iron oxide facility.

Major producers of nitrobenzene in Western Europe are shown in Table 5-15 (4). In addition, smaller producers include Hoechst in West Germany,

TABLE 5-14 Major U.S. Producers of Nitrobenzene

Producer	Location	Estimated Capacity mm lb/yr	Estimated Capacity m MT/yr
DuPont	Beaumont, Texas	350	159
	Gibbstown, New Jersey	340	154
First Chemical	Pascagoula, Mississippi	335	152
Mobay	New Martinsville, W. Virginia	200	91
Rubicon	Geismar, Louisiana	375	170
	TOTAL	1600	726

TABLE 5-15 Major Western European Producers of Nitrobenzene

Producer	Location	Estimated Capacity mm lb/yr	Estimated Capacity m MT/yr
BASF	Antwerp, Belgium	220	100
Bayer	Antwerp, Belgium	220	100
	Leverkusen, W. Germany	529	240
CdF Chimie	Saint-Avold, France	77	35
ICI	Huddersfield, U.K.	353	160
QUIMIGAL	Estarreja, Portugal	154	70
	TOTAL	1553	705

ACNA in Italy, and others in Spain, Switzerland, and the United Kingdom. Producers in Japan include Mitsubishi Chemical Industries, Mitsui Toatsu Chemicals, Sumitomo Chemical, and Hodogaya Chemicals.

Economics

Production costs of nitrobenzene for a typical installation are as follows:

Capacity: 160 mm lb/yr (73 m MT/yr)
Capital cost: BLCC, $17 mm; OSBL, $8 mm; WC, $13 mm

	¢/lb	$/MT	%
Raw materials[a]	16.2	357	68
Utilities	3.6	79	15
Operating costs	1.0	22	4
Overhead costs	3.0	66	13
Cost of production	23.8	524	100
Transfer price	28.5	628	

[a] Benzene at 17¢/lb; nitric acid at 10¢/lb

ANILINE

Aniline has one major growth market—methane diisocyanate (MDI)—which is used in the production of rigid urethane foams and urethane elastomers. It is produced primarily from benzene via a two-stage process. In the first step benzene is nitrated to nitrobenzene; the second step involves the hydrogenation of nitrobenzene to aniline. An alternative route used by some manufacturers is a single-step ammonolysis with phenol as the feed.

Technology

Hydrogenation of nitrobenzene. Processes for the reduction of nitrobenzene to aniline have been available for well over a century. Early technology involved the use of ammonium sulfide. This technology was followed by the Bechamp process, in which iron and dilute acid were used for the reduction. The process was commercialized a few years later and formed the basis for the dyestuff industry. Batch procedures were used initially but have been largely replaced by continuous-flow catalytic reduction processes for the modern, larger installations. Much of the development occurred during World War II, hastened by the availability of large volumes of low-cost hydrogen. Advantages of the catalytic technology include relatively low capital investments, lower operating and maintenance costs, and higher yields. The process occurs in the vapor phase at temperatures of up to 350°C. Yields of 98 percent are common. Among the various catalysts employed are copper compounds, sulfides of nickel, molybdenum or tungsten, and oxides of various metals.

The batch Bechamp process is carried out in a hydrochloric acid solution in the presence of cast-iron filings or powder. Instead of hydrochloric acid the aniline salt liquor produced during the reaction can be reused. Yields are generally better than 90 percent.

The reaction is as follows:

$$C_6H_5NO_2 + 3H_2 \longrightarrow C_6H_5NH_2 + 2H_2O$$

The reaction is highly exothermic, and heat removal is an important design consideration in the process. Catalytic activity is reduced over a period of time by deposition of organic material on the catalyst, and regeneration of the catalyst is required. A block flowsheet for the process is shown in Figure 5-6.

Amination. For some years Dow Chemical had produced aniline by reacting chlorobenzenes and ammonia in the presence of a copper compound. The process was actually discontinued almost 20 years ago. In 1982, USS Chem-

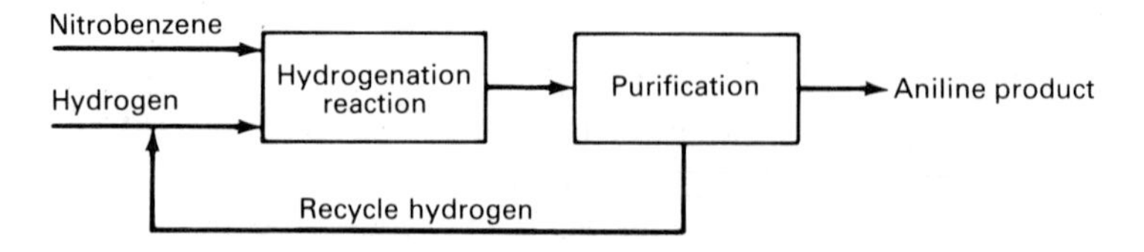

Figure 5.6 Production of aniline

icals constructed a 200 million pound per year facility at Haverhill, Ohio, which uses phenol rather than benzene as feedstock (3). This facility was the first in the United States and uses technology developed by Halcon. The process based on the ammonolysis of phenol was initially commercialized by Mitsui Petrochemical of Japan and was used at its facilities at Chiba, Japan. The one-step, reversible reaction is as follows:

$$C_6H_5OH + NH_3 \rightleftharpoons C_6H_5NH_2 + H_2O$$

The reaction is favored by high ammonia-to-phenol ratios and low reaction temperatures. The forward reaction is slightly exothermic, and reaction temperatures are relatively easy to control. The reaction occurs in the vapor phase and uses a Lewis-acid-type catalyst. The purified product is obtained by conventional and azeotropic distillation, and ammonia, as well as unreacted phenol, is recycled. Proponents of this technology claim several advantages over the traditional nitrobenzene process. It is a one-step operation, and ammonia and phenol are easily transported articles of commerce. Nitrobenzene, on the other hand, is generally produced on site by the aniline manufacturer. Capital investment is claimed to be lower, and acid regeneration is avoided.

Market

Hundreds of chemical products and intermediates are produced from aniline, with the major uses in the polymer, rubber, agricultural, and dye industries. Estimated consumption in 1983 was close to 665 million pounds in the United States, with about 60 percent going for MDI and polymeric MDI for ultimate use in the production of polyurethanes. Another 18 percent was consumed by rubber chemicals, with the balance going to drugs, pesticides, dyes, and hydroquinone. The MDI is the major growth market. Automobile manufacturers are using molded MDI exteriors parts, and this use is expected to expand. Capacity is likely to be adequate for some time, despite the shutdown of American Cyanamid's facility in West Virginia at the end of 1982. This reduction was more than compensated for by the start-up of USS Chemicals' facility at Haverhill, Ohio. U.S. producers and their approximate capacities are shown in Table 5-16 (1,2).

Although DuPont is the largest producer of aniline in the United States, it is the most subject to market fluctuations, since it absorbs only a small portion of the production. In contrast ICI and Mobay have substantially captive uses for their production. They have shut down about 90 million pounds per year of capacity. Aniline is used primarily for MDI production by Mobay and for rubber chemicals by ICI.

Table 5-17 lists the major aniline producers in Western Europe (6). In addition, a few small producers are in Switzerland and Spain. Major producers in Japan include Mitsui Petrochemicals, Mitsui Toatsu Chemicals, and Sumitomo Chemical.

Because of the major use of aniline for isocyanates used in rigid polyurethane foam manufacture, the market will substantially follow the fortunes of the construction and transportation insulation applications. The second major use of aniline is in the rubber industry, primarily for tires. This application is therefore again dependent on the state of the automobile market. Radial tires use more aniline-derived chemicals than do conventional tires, which is a positive influence on the market growth.

TABLE 5-16 Major U.S. Producers of Aniline

		Estimated Capacity		
Producer	Location	mm lb/yr	m MT/yr	Feedstock
DuPont	Beaumont, Texas	260	118	Nitrobenzene
	Gibbstown, New Jersey	170	77	Nitrobenzene
First Chem.	Pascagoula, Mississippi	250	113	Nitrobenzene
Rubicon	Geismar, Louisiana	280	127	Nitrobenzene
Mobay	New Martinsville, W. Virginia	40	18	Nitrobenzene
USS Chemicals	Haverhill, Ohio	200	91	Phenol
	TOTAL	1200	544	

TABLE 5-17 Major Western European Producers of Aniline

		Estimated Capacity	
Producer	Location	mm lb/yr	m MT/yr
BASF	Antwerp, Belgium	66	30
	Ludwigshafen, W. Germany	55	25
Bayer	Antwerp, Belgium	159	72
	Krefeld, W. Germany	386	175
ICI	Wilton, U.K.	254	115
QUIMIGAL	Estarreja, Portugal	110	50
	TOTAL	1030	467

Economics

Production costs for a typical aniline facility via nitrobenzene are as follows:

Capacity: 100 mm lb/yr (45 m MT/yr)
Capital cost: BLCC, $29 mm; OSBL, $14 mm; WC, $12 mm

	¢/lb	$/MT	%
Raw materials[a]	25.7	567	71
Utilities	1.7	37	5
Operating costs	1.9	42	5
Overhead costs	7.1	157	19
Cost of production	36.4	803	100
Transfer price	49.3	1087	

[a] Benzene at 17¢/lb; nitric acid at 10¢/lb; hydrogen at 50¢/lb

POLYISOCYANATES

The wide acceptance of polyurethane products for a variety of applications has resulted in isocyanates becoming a rapidly growing major petrochemical derivative. The major isocyanate products are toluene diisocyanate (TDI) and polymeric isocyanates. TDI is discussed in Chapter 6. The market is quite diverse, and there are many aromatic and aliphatic isocyanates. The more important of these products are diphenyl methane-4,4′ diisocyanate (MDI), hydrogenated MDI, hexylmethylene diisocyanate, and isophorone diisocyanate. Current technology involves phosgenation of a diamine to the diisocyanate, but substantial effort is being expended to develop a process route that does not involve phosgene.

Technology

Although TDI has conventionally been the prime isocyanate for flexible foam production, polymeric isocyanates have been dominant for rigid foam applications. The rigid foam market is expected to grow at a faster rate. At present polyisocyanates are produced by the phosgenation of polyamines, which are produced by the condensation of aniline with formaldehyde. The reaction is as follows:

$$C_6H_5NH_2 + HCHO \longrightarrow H_2N{-}C_6H_4{-}CH_2{-}C_6H_4{-}NH_2 + \text{oligomers}$$

The reaction product consists of a mixture of oligomeric materials, whose composition is a function of the aniline/formaldehyde mole ratio. A ratio of more than 2 primarily produces p,p'-methylenephenylenediamine (MDA). The ratio generally used will result in a polyisocyanate with a functionality of approximately 2.7. In practice pure material is distilled from the phosgenated mixture of diamines. Polyisocyanates are produced by phosgenation of the corresponding amines in monochlorobenzene. About 200 percent excess phosgene is required, and the reaction is generally carried out at three different temperatures, from 50 to 140°C. Production of the polyisocyanates is generally considered to involve a multistep reaction in which intermediate products are formed. Diphenylmethane diisocyanate (MDI) is produced together with its oligomers, the polymeric isocyanates. To produce MDI, one can either separate the diamine precursor from the higher amines and phosgenate or separate the MDI from the mixed isocyanates. The following simplified reaction shows the phosgenation:

$$H_2N{-}C_6H_4{-}CH_2{-}C_6H_4{-}NH_2 + 2COCl_2 \longrightarrow OCN{-}C_6H_4{-}CH_2{-}C_6H_4{-}NCO + 4HCl$$

A block flow diagram for a typical process is shown in Figure 5-7.

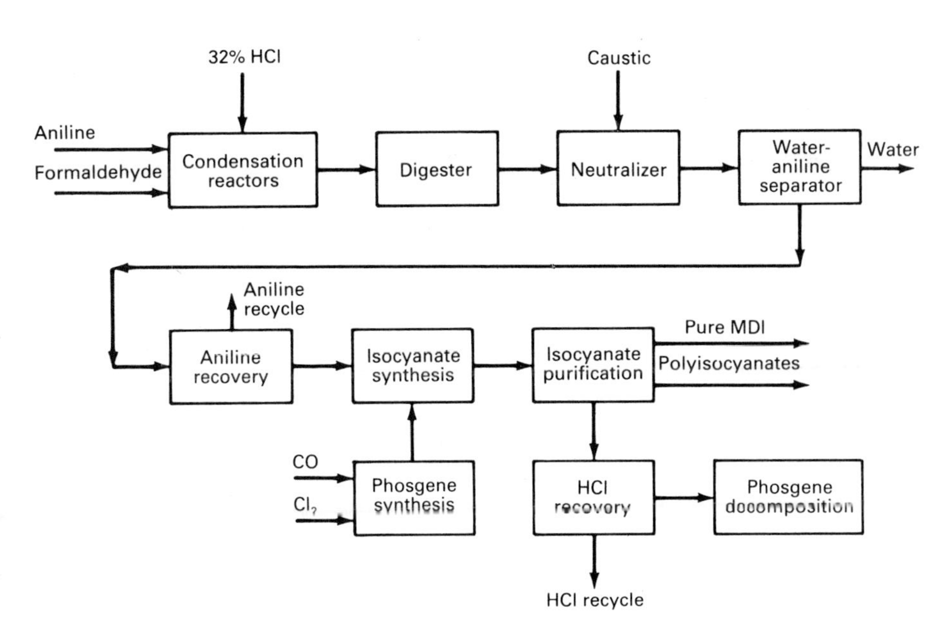

Figure 5.7 Production of polyisocyanates from aniline

The latest route to polymeric isocyanates has been developed by Arco and is based on direct carbonylation of nitrobenzene, thus eliminating the production of polyamines. The process is based on a selenium catalyst. The process produces a urethane in the initial nitrobenzene carbonylation step. The urethane is condensed with formaldehyde using a strong acid catalyst and is finally pyrolyzed to the polyisocyanate and an alcohol, which is recycled.

Market

The prime outlets for rigid polyurethanes are in insulation used in construction, appliance, transportation, and pipeline and tank applications. In addition, molded rigid polyurethane is used for decorative and structural furniture components. A prime limitation in the use of polyurethane foam in construction applications has been the flame and smoke characteristics of the products, and flame retardants are often included in the formulation. Urethanes are increasingly being used for parts in the automotive industry to take advantage of their low weight and ability to spring back to their original shape after impact. A reaction injection molding process (RIM) has been efficiently used to make elastomer-based auto parts. In many cases glass fibers are being incorporated to add both strength and stiffness to the polyurethane part.

Table 5-18 indicates the major producers of MDI and polymeric isocyanates in the United States (2,4). Production in the United States in 1983 was over

TABLE 5-18 Major U.S. Producers of MDI and Polymeric Isocyanate

Producer	Location	Estimated Capacity	
		mm lb/yr	m MT/yr
BASF Wyandotte	Geismar, Louisiana	150	68
Rubicon	Geismar, Louisiana	100	45
Mobay	Cedar Bayou, Texas	200	91
	New Martinsville, W. Virginia	100	45
Upjohn	LaPorte, Texas	250	113
	TOTAL	800	362

500 million pounds, rising to over 600 million pounds in 1984. Projections indicate a demand of about 750 million pounds by 1988. The current proportions of polymeric MDI and pure MDI used for the different end products are shown in the following table (4):

End Product	% Used
Polymeric MDI	
Laminate board	26
Spray foam	9

End Product	% Used
Polymeric MDI (continued)	
Panelling	9
Packaging	4
Other rigid foam	6
Flexible foam	12
Pure MDI	
RIM products	7
Other	6
Export	21

Table 5-19 indicates the major producers of MDI in Western Europe (3). BASF, Bayer, and ICI use their own technology. Montepolimeri uses ICI technology, and Companhia Portuguesa de Isociantos uses Upjohn technology. Major Japanese producers and their estimated capacities are shown in Table 5-20.

TABLE 5-19 Major Western European Producers of MDI

		Estimated Capacity	
Producer	Location	mm lb/yr	m MT/yr
BASF	Antwerp, Belgium	88	40
Bayer-Shell Isocyanates	Antwerp, Belgium	57	26
Bayer	Krefeld, W. Germany	254	115
Montepolimeri	Brindisi, Italy	66	30
ICI	Rotterdam, Netherlands	88	40
Comp. Port. de Isoc.	Estarreja, Portugal	110	50
ICI	Fleetwood, U.K.	88	40
	TOTAL	751	341

TABLE 5-20 Major Japanese Producers of MDI

		Estimated Capacity	
Producer	Location	mm lb/yr	m MT/yr
Nippon Polyurethane	Tsurumi	70	32
Mitsui Toatsu Chemical	Omuta	65	29
Mitsubishi Chemical	Kurosaki	65	29
Kasei Upjohn	Mie	80	36
Sumitomo Bayer Urethane	Kikumoto	65	29
	TOTAL	345	155

Economics

Typical production costs for a facility to produce pure MDI and poly-isocyanates are as follows:

Capacity: 100 mm lb/yr (45 m MT/yr)
Capital cost: BLCC, $41 mm; OSBL, $21 mm; WC, $23 mm

	¢/lb	$/MT	%
Raw materials[a]	51.5	1136	71
Utilities	8.3	183	12
Operating costs	2.4	53	3
Overhead costs	9.9	218	14
Cost of production	72.1	1590	100
By-product credit (HCl)	(2.7)	(60)	
Net cost of production	69.4	1530	
Transfer price	88.0	1940	

[a] Aniline at 38¢/lb

CYCLOHEXANE

Large-scale cyclohexane production has been available since the fifties. Its use has followed the demands for Nylon 6 and Nylon 6/6 because most of the cyclohexane produced is consumed in the production of chemical precursors for the nylons—adipic acid for Nylon 6/6 and caprolactam for Nylon 6.

Technology

Cyclohexane is manufactured by the hydrogenation of benzene, as well as recovered from natural gasoline. A large percentage is manufactured by the hydrogenation of pure benzene, since a high degree of purity is required for the oxidation of the cyclohexane to adipic acid for the ultimate manufacture of fiber. The hydrogenation reaction is as follows:

$$C_6H_6 + 3H_2 \longrightarrow \text{(S)}$$

Many processes for the hydrogenation of benzene have been developed. Although basically similar, the variations in the processes occur in the following areas:

1. Nature of the catalyst: Most of the catalysts are based on nickel.

2. Operating conditions: Both liquid- and vapor-phase processes have been developed.
3. Reactor design: Fixed or moving bed may be used.
4. Heat removal: The reaction is highly exothermic.

Processes have been developed by Lummus, BP, Stamicarbon, UOP, and others. Not all of these processes have been commercialized. The purity of the cyclohexane obtained depends primarily on the quality of the benzene used as starting material. Phillips Petroleum is presently the only U.S. company to produce cyclohexane via distillation. Cyclohexane is present in very small quantities in all crude petroleum—from about 0.05 to 0.5 percent. In distillation, cyclohexane is always accompanied by its isomer methylcyclopentane and by benzene. Normal boiling points of cyclohexane and benzene differ by less than a centigrade degree and are about 9 degrees higher than that of methylcyclopentane. Phillips reportedly combines distillation with a stage where methylcyclopentane is isomerized to cyclohexane. In addition, extractive distillation techniques can be used to obtain suitably pure cyclohexane. A block flow diagram for a typical process is shown in Figure 5-8.

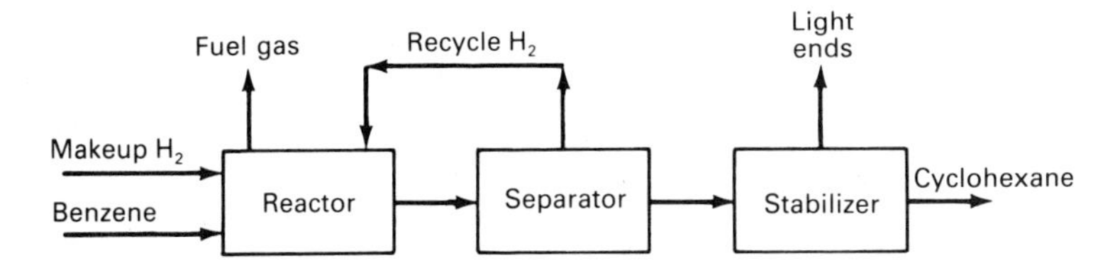

Figure 5.8 Production of cyclohexane by hydrogenation of benzene

Market

Because of its predominate use in nylons, the market for cyclohexane closely follows that for the nylons. The major applications for the nylons are for fibers used in carpets and rugs, tire cord, and knitted clothing. In the early eighties the cyclohexane industry was devastated because of the depressed state of the housing and the automobile industries. At present there is an excess of cyclohexane capacity. U.S. demand for cyclohexane for 1982 and 1983 hovered around the 200 million gallon mark, with capacity at almost double that amount. Projections are for a demand of about 230 million gallons in 1987, and there is more than ample capacity for the future (3). Current U.S. producers of cyclohexane and their estimated capacities are shown in Table 5-21 (2,3).

Exxon has a 50 million gallon per year facility at Baytown, Texas, which is on standby, and Phillips closed its 100 million gallon per year facility at

TABLE 5-21 Major U.S. Producers of Cyclohexane

Producer	Location	Estimated Capacity		
		mm gal/yr	mm lb/yr	m MT/yr
Champlin	Corpus Christi, Texas	22	143	65
DuPont	Corpus Christi, Texas	50	325	147
Gulf	Port Arthur, Texas	38	247	112
Phillips	Borger, Texas	45	292	132
	Guayama, Puerto Rico	89	578	262
Sun	Tulsa, Oklahoma	28	182	82
Texaco	Port Arthur, Texas	60	389	177
Union Oil	Beaumont, Texas	30	195	88
	TOTAL	362	2351	1065

Sweeney, Texas, in 1984. About 80 percent of the output ends up in nylon production: 54 percent for Nylon 6/6 and 26 percent for Nylon 6. Most of the remainder is for the export market, which is expected to decline steadily as the world builds capacity. Cyclohexane consumption is clearly dominated by adipic acid for fibers, plastics, and lubricants. Most of adipic acid production is based on cyclohexane, with the remainder based on phenol.

Major Western European producers of cyclohexane are shown in Table 5-22 (4). Major Japanese producers are shown in Table 5-23.

TABLE 5-22 Major Western European Producers of Cyclohexane

Producer	Location	Estimated Capacity		
		mm gal/yr	mm lb/yr	m MT/yr
Petrochim	Antwerp, Belgium	36	232	105
CdF Chimie	Carling, France	17	110	50
Elf-Aquitaine	Laco, France	8	55	25
Ruhr Oel	Gelsenkirchen, W. Germany	34	221	100
Wintershall	Lingen, W. Germany	34	221	100
Mobil Oil	Napoli, Italy	14	88	40
Esso Chemie	Rotterdam, Netherlands	66	428	194
ERT	La Rabida, Spain	27	176	80
ICI	Middlesbrough, U.K.	102	662	300
	TOTAL	338	2193	994

TABLE 5-23 Major Japanese Producers of Cyclohexane

Producer	Location	Estimated Capacity		
		mm gal/yr	mm lb/yr	m MT/yr
Mitsubishi Chemical	Mizushima	34	220	100
Idemitsu Petrochemical	Tokuyama	31	200	91
Kanto Denka Kogyo	Shibukawa, Mizushima	8	50	23

TABLE 5-23 Major Japanese Producers of Cyclohexane (*continued*)

Producer	Location		Estimated Capacity		
			mm gal/yr	mm lb/yr	m MT/yr
Nippon Steel	Hirohata		12	80	36
Sumitomo Chemical	Niihama		17	110	50
Toray	Kawasaki		51	330	150
Ube Industries	Ube, Sakai		58	375	170
		TOTAL	211	1365	620

Economics

Typical production costs of cyclohexane via the hydrogenation of benzene are as follows:

Capacity: 400 mm lb/yr (181 m MT/yr)
Capital cost: BLCC, $11 mm; OSBL, $5 mm; WC, $28 mm

	¢/lb	$/MT	%
Raw materials[a]	18.9	417	89
Utilities	0.6	13	3
Operating costs	0.3	7	1
Overhead costs	1.5	33	7
Cost of production	21.3	470	100
Transfer price	22.5	496	

[a] Benzene at 17¢/lb

ADIPIC ACID

Over 90 percent of all adipic acid produced in the United States is used in the production of Nylon 6/6 fibers and resins by reaction with hexamethylene diamine (HMDA). In the United States adipic acid is primarily produced by the catalytic air oxidation of cyclohexane, followed by nitric air oxidation. With the exception of Allied, which produces small quantities of adipic acid as a by-product of its caprolactam production, all producers use this process and have captive requirements for the manufacture of Nylon 6/6.

Technology

The predominate technology in adipic acid production involves a two-stage liquid-phase oxidation of cyclohexane. The cyclohexane is conventionally derived directly from petroleum refinery operations or, as an alternative, by hydrogenation of benzene. The hydrogenation, usually involving a Raney nickel catalyst, is virtually quantitative and proceeds according to the following reaction:

$$C_6H_6 + 3H_2 \longrightarrow C_6H_{12}$$

To produce a fiber-grade cyclohexane product, high-purity benzene is required. Further details are given in the previous section.

The first stage of oxidation involves the conversion of cyclohexane to a mixture of cyclohexanol and cyclohexanone. This mixture is commonly called KA (ketone-alcohol) oil. The reaction is typically carried out in the presence of metaboric acid, which reduces excessive oxidation and increases yields. Operating conditions are approximately 160°C and 10 atmospheres, employing a cobalt naphthenate catalyst. Since conversion is only about 10 percent per pass, the KA oil must be separated and the unconverted cyclohexane recycled. The reaction follows:

$$C_6H_{12} + O_2 \longrightarrow C_6H_{10}{=}O + C_6H_{11}OH$$

The second stage of oxidation with nitric acid, at approximately 80°C and 3 atmospheres, produces adipic acid. Yields, based on the cyclohexane, are approximately 80 percent of theoretical. The second oxidation reaction is as follows:

$$C_6H_{10}{=}O + C_6H_{11}OH \xrightarrow{HNO_3} HOOC(CH_2)_4COOH$$

A typical block flow diagram, including the hydrogenation of benzene to cyclohexane, is shown in Figure 5-9.

The cost of raw materials accounts for almost 60 percent of the cost of production of the adipic acid—the major items being the cyclohexane and the nitric acid. The increasing costs for feedstock have led to efforts to develop alternative routes to adipic acid. One route is based on propylene and involves the oxidation of propylene to acrylic acid or its methyl ester, followed by catalytic coupling, hydrogenation, and hydrolysis to adipic acid. Most of the research involves the coupling step. A second route, based on butadiene, is being studied by BASF and involves a two-step carbonylation (4,5). The butadiene is reacted with carbon monoxide and methanol at high pressures of up to about 600 atmospheres and at 120°C using a cobalt catalyst system. In this step methyl-3-pentenoate is formed. In the second stage this compound is further reacted

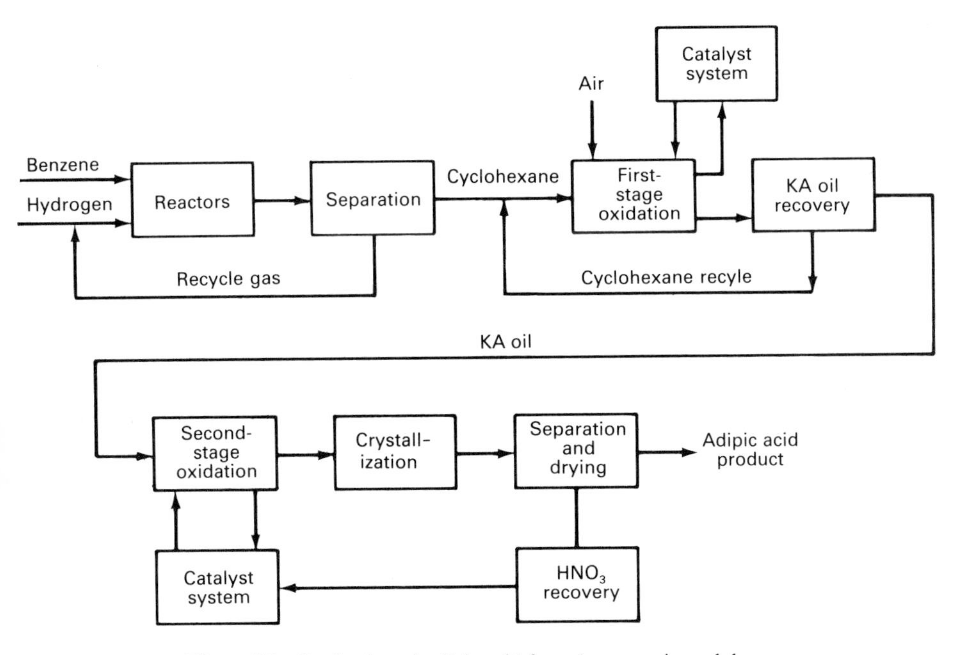

Figure 5.9 Production of adipic acid from benzene via cyclohexane

with carbon monoxide at 30 atmospheres and 185°C to produce dimethyladipate and isomers. After separation of the isomers, hydrolysis yields the adipic acid.

Market

As in the case of HMDA, the market for adipic acid closely follows that for Nylon 6/6. Nylon 6/6 fibers are used primarily for carpets, clothing, and tire cord and were hard hit in the early eighties but recovered markedly by 1983. Current U.S. producers are shown in Table 5-24 (2,3).

TABLE 5-24 Current U.S. Producers of Adipic Acid

			Estimated Capacity	
Producer	Location		mm lb/yr	m MT/yr
Allied	Hopewell, Virginia		30	14
DuPont	Orange, Texas		400	181
	Victoria, Texas		700	317
Monsanto	Pensacola, Florida		600	272
		TOTAL	1730	784

TABLE 5-25 Major Western European Producers of Adipic Acid

Producer	Location	Estimated Capacity mm lb/yr	m MT/yr	Raw Material
UCB-Ftal	Oostende, Belgium	55	25	Phenol
Rhone-Poulenc	Chalampe, France	507	230	Cyclohexane
BASF	Ludwigshafen, W. Germany	441	200	Cyclohexane
Bayer	Leverkusen, W. Germany	88	40	Cyclohexane
Montedison	Novara, Italy	110	50	Phenol
ICI	Wilton, U.K.	662	300	Cyclohexane
	TOTAL	1863	853	

U.S. demand in 1983 was over 1.1 billion pounds and is projected to grow to almost 1.3 billion pounds in 1987 (3). The second major end use, after Nylon 6/6 fibers and resins, is for production of hexamethylenediamine, which is a coreactant for Nylon 6/6. Other uses are for plasticizers and polyurethane resins. As can be seen from the table, substantial overcapacity exists and new facilities will probably not be required for many years. In addition, imports from European producers amount to about 4 percent of U.S. demand and pose a threat to domestic producers.

Major producers of adipic acid in Western Europe are shown in Table 5-25 (6). Major Japanese producers of adipic acid are Asahi Chemical Industries, Sumitomo Chemical, Ube Industries, Kanto Denka Kogyo, and Honshu Chemical Industries.

Economics

Typical production costs for adipic acid are as follows:

Capacity: 200 mm lb/yr (91 m MT/yr)
Capital cost: BLCC, $55 mm; OSBL, $58 mm; WC, $25 mm

	¢/lb	$/MT	%
Raw materials[a]	22.6	498	59
Utilities	5.5	121	14
Operating costs	2.1	46	6
Overhead costs	7.8	172	21
Cost of production	38.0	837	100
Transfer price	55.0	1213	

[a] Cyclohexane at 20¢/lb; nitric acid at 10¢/lb

CUMENE, PHENOL, AND ACETONE

Phenol is the starting material for a variety of important petrochemical and plastic products. Many processes are possible for its manufacture and many have been employed commercially in the past. Cumene is generally the feedstock of choice, and acetone is a major by-product of the most common process. Some of the end uses of phenol face competition from substitute materials that are not derived from phenol. Phenol-based adhesives are used extensively in the manufacture of plywood and are therefore highly dependent on the housing market.

Technology

The major markets for phenol are the production of phenolic resins and bisphenol-A, which require the use of phenol. Other market products can be made from alternative raw materials. For example, caprolactam and adipic acid can be produced from phenol or from cyclohexane. In the United States, Western Europe, and Japan, the production of phenol from benzene and propylene via cumene accounts for the major share of current production. A smaller fraction is primarily divided among conversion from toluene via benzoic acid, from benzene via chlorobenzene, and from benzene via benzene sulfonate. A small amount is obtained from coal tar.

The alkylation of benzene with propylene to produce cumene is a well-established technology, similar to that for the production of ethylbenzene. The alkylation generally uses an aluminum chloride catalyst. Pure cumene is obtained by distillation, and unconverted reactants are recycled to improve yields. The reaction follows:

$$C_6H_6 + CH_3CH{=}CH_2 \longrightarrow C_6H_5CH(CH_3)_2$$

The reaction is exothermic and is generally carried out in the vapor phase at about 40 atmospheres of pressure and about 220°C.

The conversion of cumene to phenol was developed over 35 years ago by Hercules and Distillers (presently BP Chemicals), and the first facility to use the process was built in Montreal, Canada, in the early fifties by Shawinigan (presently Gulf Oil Canada). The process involves the following steps:

1. Oxidation of cumene with air to produce cumene hydroperoxide.
2. Separation and recycling of unreacted cumene from the hydroperoxide.
3. Cleavage of the cumene hydroperoxide to phenol and acetone.
4. Separation of pure phenol and acetone products.

The reactions are as follows:

$$C_6H_5\text{—}\underset{\displaystyle CH_3\;\;CH_3}{CH} + O_2 \longrightarrow C_6H_5\text{—}\underset{\displaystyle CH_3\;\;CH_3}{COOH}$$

$$C_6H_5\text{—}\underset{\displaystyle CH_3\;\;CH_3}{COOH} \longrightarrow C_6H_5\text{—}OH + CH_3\overset{\overset{\displaystyle O}{\|}}{C}CH_3$$

The cleavage of the cumene hydroperoxide to phenol and acetone must be conducted at moderate temperatures, since thermal decomposition produces undesirable by-products. Yields are over 95 percent for each of the steps. Figure 5-10 represents a block flow diagram for the process.

Since about 0.6 pound of acetone is produced for each pound of phenol, the market for the by-product acetone has a considerable effect on the economics of the process. A large part of acetone demand is due to its use as a raw material

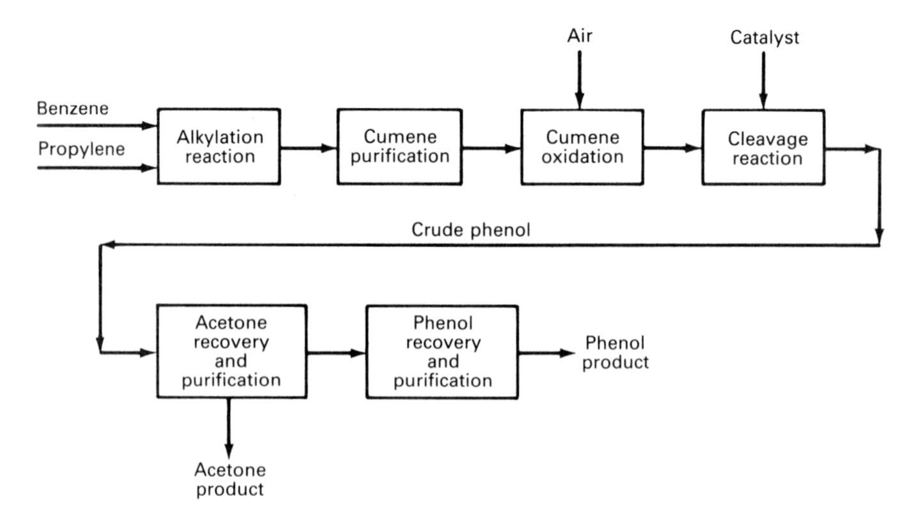

Figure 5.10 Production of phenol from benzene and propylene via cumene

in the production of methyl methacrylate via the acetone cyanohydrin process. Research in new processes for the production of methyl methacrylate is quite active, and developments could have a major effect on the economics for the production of phenol via cumene oxidation. Because of this research, there is much interest in the development of processes for the production of phenol without an acetone by-product. One such process has been developed by CE Lummus (1). It involves oxidation of toluene to benzoic acid, followed by further oxidation to phenol over a novel copper-bearing catalyst. Many commercial processes are available for the oxidation of toluene to benzoic acid, usually carried out in the liquid phase using a cobalt catalyst.

Market

Phenol is a fairly typical large-volume organic chemical, moving up and down with economic cycles. In the United States the downturn in 1982 was followed by an increase of 14 percent in 1983 to a level of over 2.5 billion pounds (2). This production level is considerably lower than the installed capacity of over 3.5 billion pounds per year. The 1982 production figures were down about 15 percent from 1981 and 24 percent from the record set in 1979. Projected demand for 1988 is about 3.1 billion pounds (4). The demand for phenol is largely dependent on the housing industry as the end use for its various derivatives. It is a starting material for plywood adhesives, carpet fibers, and other polymeric materials. Automotive and other transportation products are another important end use for phenol derivatives such as bisphenol-A. Major end uses as a percentage of phenol production are shown in the following table (4):

End Use	%
Phenolic resins	45
Bisphenol-A	20
Caprolactam	13

Bisphenol-A is an intermediate for both polycarbonates and epoxy resins. Major producers in the United States for cumene and phenol are shown in Tables 5-26 and 5-27, respectively (3,5,9). Tables 5-28 and 5-29 show major Western European producers of cumene and phenol, respectively (11). All of the Western European producers shown in Tables 5-28 and 5-29 use cumene feedstock with the exception of Chemische Industrie Rijnmond in the Netherlands, which uses toluene. In addition, several producers in Western Europe manufacture phenol from coal tar. Tables 5-30 and 5-31 show major Japanese producers of cumene and phenol, respectively.

TABLE 5-26 Major U.S. Producers of Cumene

Producer	Location	Estimated Capacity mm lb/yr	Estimated Capacity m MT/yr	Use
Apex Oil	Blue Island, Illinois	110	50	Captive
Ashland Oil	Catlettsburg, Kentucky	400	181	Sold
Georgia-Pacific	Pasadena, Texas	750	340	Partially captive
Getty Oil	El Dorado, Kansas	135	61	Captive
Koch Refining	Corpus Christi, Texas	400	181	Sold
Shell Chemical	Deer Park, Texas	720	327	Captive
Amoco Chemical	Texas City, Texas	30	14	Captive
Texaco Chemical	Westville, New Jersey	140	63	Sold
Champlin Petroleum	Corpus Christi, Texas	400	181	
	TOTAL	3085	1398	

TABLE 5-27 Major U.S. Producers of Phenol

Producer	Location	Estimated Capacity mm lb/yr	Estimated Capacity m MT/yr
Allied[a]	Frankford, Pennsylvania	500	227
USS Chemical[b]	Haverhill, Ohio	520	236
Shell Chemical[c]	Deer Park, Texas	500	227
Monsanto[d]	Alvin, Texas	500	227
Georgia Pacific[e]	Plaquemine, Louisiana	330	150
	Boundbrook, New Jersey	157	71
Dow Chemical[f]	Oyster Creek, Texas	470	213
General Electric[g]	Mt. Vernon, Indiana	400	181
Other[h]		300	136
	TOTAL	3677	1668

[a] Uses phenol mainly as a precursor for Nylon 6 resins and fibers by Allied.

[b] Trying to develop downstream intermediates such as bisphenol-A and aniline using phenol feedstock.

[c] Produces bisphenol-A for merchant market.

[d] Produced phenolic resins and adipic acid for merchant market, but has withdrawn from business.

[e] Large percentage of production is used internally. The company started up a 157 million lb/yr facility at Boundbrook, New Jersey, in 1984.

[f] Supplies merchant market, but captive use primarily for polycarbonates is growing.

[g] Uses phenols primarily for polycarbonates internally.

[h] Includes Kalama, which uses toluene feedstock. Union Carbide has a 220 million lb/yr phenol facility with associated cumene and bisphenol-A units at Penuelas, Puerto Rico, on standby.

TABLE 5-28 Major European Producers of Cumene

Producer	Location	Estimated Capacity mm lb/yr	m MT/yr
Neste Oy	Kulloo, Finland	154	70
Progil Electrochimie	Le Pont-de-Claix, France	265	120
Rhone-Poulenc	Roussillon, France	331	150
Shell Chimie	Pauillac, France	221	100
Chemische Werke Huels	Marl, W. Germany	238	108
Ruhr Oel	Gelsenkirchin, W. Germany	728	330
Anic	Gela, Italy	331	150
	Porto-Torres, Italy	331	150
Montedison	Priolo, Italy	396	180
Saras Chimica	Sarroch, Italy	441	200
Dow Chemical	Terneuzen, Netherlands	529	240
Ertisa	Huelva, Spain	265	120
BP Chemicals	Grangemouth, U.K.	221	100
ICI	Middlesbrough, U.K.	276	125
	TOTAL	4727	2143

TABLE 5-29 Major Western European Producers of Phenol

Producer	Location	Estimated Capacity mm lb/yr	m MT/yr
Neste Oy	Kulloo, Finland	110	50
Progil Electrochimie	Le Pont-de-Claix, France	198	90
Rhone-Poulenc	Roussillon, France	132	60
Phenolchemie	Gladbeck, W. Germany	882	400
Anic	Porto-Torres, Italy	198	90
	Solbiate, Italy	55	25
Montedison	Mantova, Italy	474	215
Chem. Ind. Rijnmond	Rotterdam, Netherlands	562	255
Ertisa	Huelva, Spain	187	85
BP Chemicals	Grangemouth, U.K.	106	48
ICI	Billingham, U.K.	185	84
Shell Chemicals	Ellesmere Port, U.K.	265	120
	TOTAL	3354	1522

TABLE 5-30 Major Japanese Producers of Cumene

Producer	Location	Estimated Capacity mm lb/yr	m MT/yr
Mitsubishi Petrochemical	Kashima	240	109
Mitsui Petrochemical	Chiba	490	222
Shin Daikowa	Iwakuni, Yokkaichi	440	200
	TOTAL	1170	531

TABLE 5-31 Major Japanese Producers of Phenol

Producer	Location		Estimated Capacity	
			mm lb/yr	m MT/yr
Mitsui Petrochemical	Chiba		45	20
Mitsui Toatsu Chemicals	Senpoku		220	100
Mitsubishi Petrochemical	Kashima		220	100
		TOTAL	485	220

Most of the acetone produced in the United States, Western Europe, and Japan is a by-product of phenol production by cumene peroxidation. Demand in the United States was about 1.9 billion pounds in 1983 and is projected to grow to over 2.1 billion pounds in 1988 (13). Growth is expected to be about 3 percent a year over the next few years. The percentages of the major uses for acetone in the United States are shown in the following table.

End Use	%
Methyl methacrylate and other acrylates	33
Solvents	17
Methyl isobutyl ketone	10
Bisphenol-A	9
Aldol chemicals	7

Major U.S. and Western European producers of acetone are shown in Tables 5-32 and 5-33, respectively (9,11,13).

In the United States all but about a billion pounds of acetone per year is produced as a by-product of phenol production via cumene peroxidation. The second major production category is via the dehydrogenation of isopropyl alcohol. Shell Chemical produces 700 million pounds per year, Union Carbide produces 170 million pounds per year, and Eastman produces 80 million pounds per year in this way. Arco's 40 million pound per year facility produces crude material as a by-product of propylene oxide production.

In Western Europe almost all of the acetone is produced as a by-product of phenol production, and therefore, the major facilities primarily produce phenol via cumene peroxidation, as shown in Table 5-33. Major Japanese producers of acetone other than those listed in Table 5-31 include Nippon Petrochemicals, Kanegafuchi Chemical Industry, Kyowa Hakko Kogyo, Sumitomo Chemical, and Shin-Daikyowa Petrochemical.

TABLE 5-32 Major U.S. Producers of Acetone

Producer	Location	Estimated Capacity	
		mm lb/yr	m MT/yr
Allied	Frankford, Pennsylvania	300	136
Apex Oil	Blue Island, Illinois	53	24
Arco	Bayport, Texas	40	18
Dow Chemical	Freeport, Texas	280	127
Eastman	Kingsport, Tennessee	80	36
General Electric	Mt. Vernon, Indiana	240	109
Georgia-Pacific	Boundbrook, New Jersey	95	43
	Plaquemine, Louisiana	200	91
Getty Oil	El Dorado, Kansas	57	26
Shell Chemical	Deer Park, Texas	700	317
	Wood River, Illinois	300	136
Union Carbide	Institute, W. Virginia	170	77
USS Chemicals	Haverhill, Ohio	320	145
	TOTAL	2835	1285

TABLE 5-33 Major Western European Producers of Acetone

Producer	Location	Estimated Capacity	
		mm lb/yr	m MT/yr
Neste Oy	Kulloo, Finland	66	30
Progil Electrochimie	Le Pont-de-Claix, France	119	54
Rhone-Poulenc	Roussillon, France	79	36
Shell Chimie	Berre-L'Etang, France	154	70
Deutsche Texaco	Moers, W. Germany	79	36
Phenolchemie	Gladbeck, W. Germany	551	250
Anic	Porto-Torres, Italy	121	55
	Solbiate, Italy	33	15
Montedison	Mantova, Italy	287	130
Shell Chemie	Rotterdam, Netherlands	221	100
Ertisa	Huelva, Spain	121	55
BP Chemicals	Grangemouth, U.K.	66	30
	Hull, U.K.	110	50
ICI	Billingham, U.K.	110	50
Shell Chemicals	Ellesmere Port, U.K.	258	117
	TOTAL	2375	1078

Economics

Production costs for cumene and phenol are shown in the following tables for a typical facility. In both cases the starting point is benzene and propylene.

Cumene

Capacity: 400 mm lb/yr (181 m MT/yr)
Capital cost: BLCC, $18 mm; OSBL, $9 mm; WC, $31 mm

	¢/lb	$/MT	%
Raw materials[a]	20.0	441	86
Utilities	1.2	26	5
Operating costs	0.3	7	1
Overhead costs	1.8	40	8
Cost of production	23.3	514	100
By-product credit (steam)	1.4	31	
Net cost of production	21.9	483	
Transfer price	23.9	527	

[a] Benzene at 17¢/lb; propylene at 20¢/lb

Phenol and Acetone

Capacity: 400 mm lb/yr (181 m MT/yr)
Capital cost: BLCC, $103 mm; OSBL, $50 mm; WC, $35 mm

	¢/lb	$/MT	%
Raw materials[a]	28.4	626	68
Utilities	6.8	150	16
Operating costs	1.3	29	3
Overhead costs	5.1	112	13
Cost of production	41.6	917	100
By-product credit[b]	15.3	337	
Net cost of production	26.3	580	
Transfer price	37.8	833	

[a] Benzene at 17¢/lb; propylene at 20¢/lb
[b] Acetone at 24¢/lb

CAPROLACTAM

Virtually all of the caprolactam produced is used in the production of Nylon 6 fibers, resins, and film. The end use markets are generally considered to be mature and are related to population growth and housing starts.

Technology

Many production routes lead to caprolactam, but most are based on one of three raw materials: cyclohexane, phenol, or toluene. At present the Beckmann synthesis, which uses cyclohexanone as an intermediate, is the commercially dominant route. Cyclohexanone is generally produced by the liquid-phase air oxidation of cyclohexane, which produces a mixture of cyclohexanol and cyclohexanone. The cyclohexanol is generally separated and dehydrogenated to produce additional cyclohexanone. The cyclohexanone then goes through the following reactions:

$$\text{Cyclohexanone} + [H_2NOH{\cdot}H_2SO_4] \longrightarrow \text{Cyclohexanone oxime} + (NH_4)_2SO_4 + H_2O$$

$$\text{Cyclohexanone oxime} + [H_2SO_4{\cdot}SO_3] \longrightarrow C_6H_{10}O{\cdot}NH{\cdot}H_2SO_4$$

$$C_6H_{10}O{\cdot}NH{\cdot}H_2SO_4 \xrightarrow{\text{aqueous } NH_3} \text{Caprolactam (NH)} + (NH_4)_2SO_4$$

Cyclohexanone

Cyclohexanone oxime

Caprolactam

The process involves the oximation of the cyclohexanone, followed by rearrangement to the lactam via strong acid catalysts. The hydroxylamine sulfate can be obtained by reaction of an equimolar mixture of ammonium hydroxide and ammonium nitrite and sulfur dioxide. Hydrolysis followed by neutralization with sulfuric acid yields the hydroxylamine sulfate and ammonium sulfate. Formation of the cyclohexanone oxime yields additional ammonium sulfate. Contact with oleum results in the nearly quantitative rearrangement to caprolactam. In the neutralization of the final product, additional ammonium sulfate is produced. As many as 4 moles of ammonium sulfate are produced for each mole of caprolactam. Because ammonium sulfate is a low-value by product, modifications of the process have aimed at a reduction of the amount produced. Figure 5-11 represents a block flow diagram of the process.

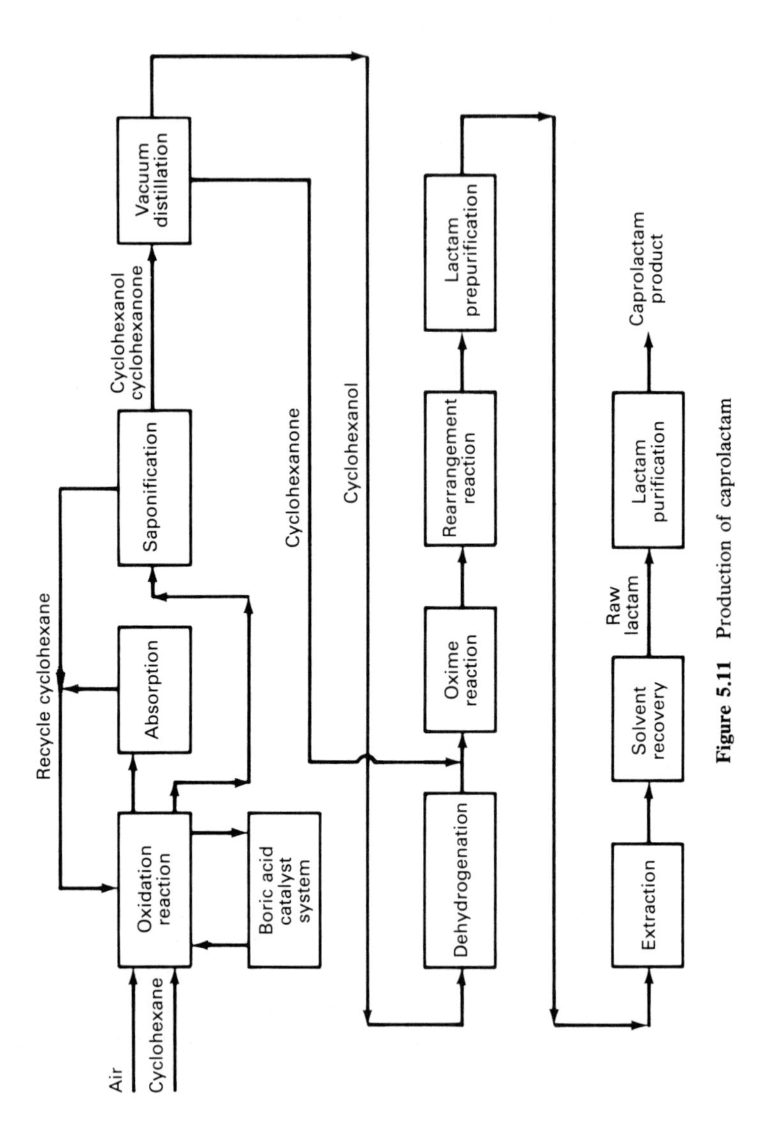

Figure 5.11 Production of caprolactam

TABLE 5-34 U.S. Producers of Caprolactam

Producer	Location	Estimated Capacity mm lb/yr	Estimated Capacity m MT/yr	Raw Material Used
Allied	Hopewell, Virginia	510	231	Phenol
Badische	Freeport, Texas	350	159	Cyclohexane
Nipro	Augusta, Georgia	360	163	Cyclohexane
	TOTAL	1220	553	

Market

Since almost 90 percent of the caprolactam produced in the United States is used in the production of Nylon 6 fibers, demand is highly dependent on the housing and transportation markets, where much of the fiber is used. In the early eighties the depressed state of the automobile markets limited the demand for the nylon in seat belts, hoses, and industrial tire cord. The carpet industry was also depressed. The mid-eighties saw strong growth in Nylon 6 fiber consumption, and by 1984 producers were operating at close to full capacity. Demand in 1982 was close to 800 million pounds in the United States and increased to almost 840 million pounds in 1983. Figures for 1984 were expected to exceed the record of 950 million pounds set in 1979 (2,4).

Installed capacity of caprolactam in the United States is in excess of 1.2 billion pounds per year, divided among three producers, as shown in Table 5-34. Nipro's production is for the merchant market, whereas Allied and Badische use part of their output for the production of Nylon 6.

Major caprolactam producers in Western Europe are shown in Table 5-35 (6). Major producers in Japan include Nippon Laectam, Mitsubishi, Toyo Rayon, and Ube. All use cyclohexane feedstock.

TABLE 5-35 Major Western European Producers of Caprolactam

Producer	Location	Estimated Capacity mm lb/yr	Estimated Capacity m MT/yr	Raw Material
BASF	Antwerp, Belgium	331	150	Cyclohexanone
	Ludwigshafen, W. Germany	309	140	Cyclohexane
Bayer	Antwerp, Belgium	220	100	Cyclohexane
Montedison	Porto-Marghera, Italy	225	102	Phenol
Soc. Chim. Dauna	Mont-Santangelo, Italy	176	80	Toluene
Soc. Chim. del Friuli	Torviscosa, Italy	35	16	Toluene
DSM	Geleen, Netherlands	485	220	Phenol
PROQUIMED	El Grao-de Castellon de la Plana, Italy	88	40	Cyclohexane
	TOTAL	1869	848	

Economics

Typical production costs for caprolactam from cyclohexane are as follows:

Capacity: 200 mm lb/yr (91 m MT/yr)
Capital cost: BLCC, $150 mm; OSBL, $69 mm; WC, $45 mm

	¢/lb	$/MT	%
Raw materials[a]	40.6	895	55
Utilities	12.7	280	17
Operating costs	4.3	95	6
Overhead costs	16.9	373	22
Cost of production	74.5	1643	100
By-product credit[b]	(7.0)	(154)	
Net cost of production	67.5	1489	
Transfer price	100.4	2214	

[a] Cyclohexane at 20¢/lb; ammonia at 10¢/lb; oleum at 4¢/lb
[b] Ammonium sulfate at 4¢/lb

NYLON FIBERS

Nylon 6/6, which is the condensation product of adipic acid and hexamethylene diamine, accounts for most of the production of nylon in the United States. Nylon 6, produced from caprolactam, is the second major nylon produced outside the United States. Probably the production locations differ because Nylon 6/6 was developed in the United States by DuPont in the late thirties and Nylon 6 by BASF in Germany at about the same time.

Technology

Nylon 6/6 fiber is produced by melt spinning nylon polymer produced by a two-stage condensation polymerization of hexamethylene diamine and adipic acid. In the first stage hexamethylene diamine adipate (AH salt) is produced by the following reaction:

$$\underset{\text{Adipic acid}}{HOOC(CH_2)_4COOH} + \underset{\text{HMDA}}{H_2N(CH_2)_6NH_2} \longrightarrow \underset{\text{AH salt}}{HOOC(CH_2)_4CONH(CH_2)_6NH_2} + H_2O$$

The second stage is the polycondensation reaction as follows:

$$nHOOC(CH_2)_4CONH(CH_2)_6NH_2 \longrightarrow [CO(CH_2)_4CONH(CH_2)_6NH]_n + nH_2O$$

As described previously, adipic acid is produced by a two-stage, liquid-phase air oxidation of cyclohexane. The several routes to hexamethylenediamine are described in the section on HMDA. A block flow diagram for the manufacture of Nylon 6/6 filament, including the production of cyclohexane, adipic acid, and HMDA from butadiene, is shown in Figure 5-12.

In the production of Nylon 6, caprolactam monomer is continuously polymerized in the presence of water, stabilizers, and modifying additives. For the production of chips suitable for injection molding, extrusion, film, fiber, and filament yarn, the raw polymer melt is extruded into strands and cut into granules. Monomers can be removed by extraction with hot water. After dewatering, the chips are dried. Recycling of the extracted monomer is usually involved. Both batch and continuous processes are available. A typical block flow diagram is shown in Figure 5-13.

Although Nylon 6/6 and Nylon 6 fibers have some different properties, they are essentially produced by the same spinning methods. The chips are metered, then melted, and then pumped to the spinning station. The molten polymer is fed into a metering pump; filtered and extruded through a spinneret into the quench duct, where the polymer solidifies into fibers; and drawn out by a drum. In this manner the fibers undergo a slight stretching. They are then wetted by lubricating agents and antistatic agents and collected on a bobbin.

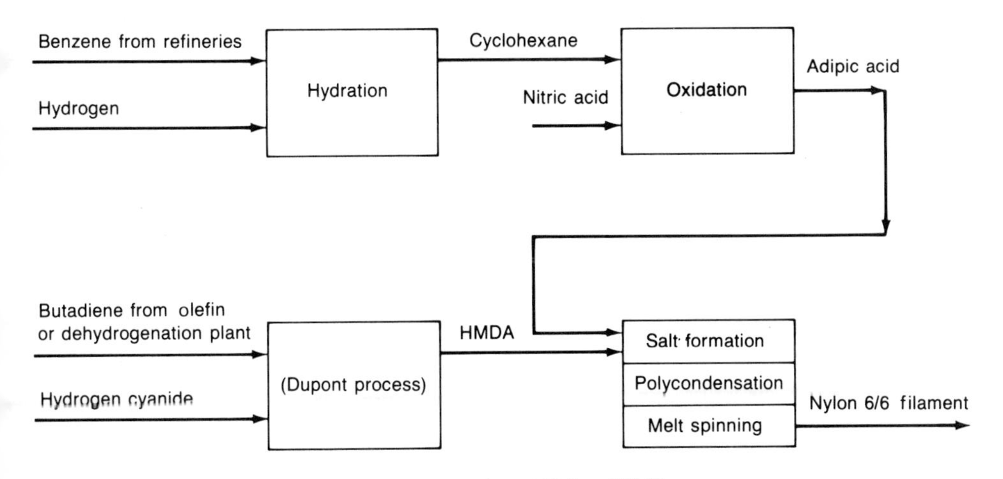

Figure 5.12 Production of Nylon 6/6 filament

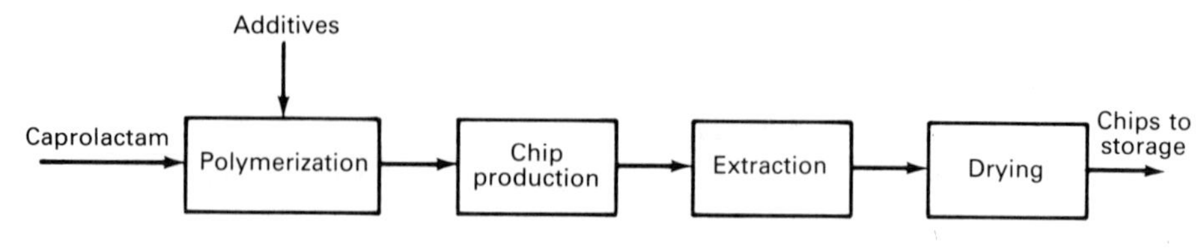

Figure 5.13 Production of Nylon 6 chips

After a storage time of several days, they are drawn and, if the product is yarn, are wound onto a tube for shipment. For the production of staple, the extruded fibers are collected to form tow and stored in a container. The differences in spinning occur because Nylon 6/6 has a higher melting point than Nylon 6. Nylon staple is often blended with other fibers, especially rayon, cotton, and wool. Nylon 6 is generally said to be easier to dye than Nylon 6/6.

Market

Table 5-36 shows the major U.S. producers of nylon (1). Table 5-37 indicates the major nylon producers in Western Europe (2). Major Japanese producers, virtually all manufacturing Nylon 6, include Teijin, Toray, Asahi Chemical, Toyobo, Kanebo and Unitika.

TABLE 5-36 Major U.S. Producers of Nylon

Producer	Location	Estimated Capacity		Product
		mm lb/yr	m MT/yr	
Akzona	Central, S. Carolina	40	18	Nylon 6
	Enka, N. Carolina	20	9	Nylon 6
	Lowland, Tennessee	160	73	Nylon 6
Allied	Chesterfield, Virginia	300	136	Nylon 6
	Irmo, S. Carolina	180	82	Nylon 6
Badische	Anderson, S. Carolina	175	79	Nylon 6
Camac	Bristol, Virginia	20	9	Nylon 6
Courtaulds	LeMoyne, Alabama	10	5	Nylon 6
DuPont	Camden, S. Carolina; Chattenooga, Tennessee; Martinsville, Richmond, and Waynesboro, Virginia; Seaford, Delware	1327	602	Nylon 6/6
Monsanto	Greenwood, S. Carolina	200	91	Nylon 6/6
	Pensacola, Florida	375	170	Nylon 6/6
Star Fibers	Edgefield, S. Carolina	25	11	Nylon 6
Wellman	Johnsonville, S. Carolina	40	18	Nylon 6/6
	TOTAL	2872	1303	

TABLE 5-37 Major Western European Producers of Nylon

Producer	Location	Estimated Capacity		Product
		mm lb/yr	m MT/yr	
Bayer	Antwerp, Belgium	22	10	Nylon 6
Beaulieu-Nylon	Kruishoutem, Belgium	26	12	Nylon 6
Fabelte Industries	Zwijnaarde, Belgium	13	6	Nylon 6/6
Rhone-Poulenc	Saint-Laurent-Blangy, France	159	72	Nylon 6/6
DuPont	Uentrop-Uber-Hamm, W. Germany	126	57	Nylon 6/6
Enka	Obenburg-am-Main, W. Germany	77	35	Nylon 6
ICI	Ostringen, W. Germany	99	45	Nylon 6/6
Norddeutsche Faser.	Neumunster, W. Germany	55	25	Nylon 6, 6/6
Rhodia	Freiburg, W. Germany	49	22	Nylon 6/6
Textil. Deggendorf	Deffendorf, W. Germany	18	8	Nylon 6/6
Snia	Sligo, Ireland	18	8	Nylon 6
Wellman	Mullagh, Ireland	11	5	Nylon 6, 6/6
Anicfibre	Pisticci, Italy	44	20	Nylon 6
Aquafil	Arco, Italy	22	10	Nylon 6
Bemberg	Gozzano, Italy	18	8	Nylon 6
Radici	Gandino, Italy	40	18	Nylon 6
Snia Fibre	Castellaccio, Italy	93	42	Nylon 6
	Cesano-Maderno, Italy	35	16	Nylon 6
	Varedo, Italy	31	14	Nylon 6
Soc. It. Nailon	Vercelli, Italy	22	10	Nylon 6/6
Enka	Emmen, Netherlands	88	40	Nylon 6
Ind. Quim. Textiles	Andoain, Spain	18	8	Nylon 6
Nurel	Zaragoza, Spain	22	10	Nylon 6
La Seda de Barcel.	Alcala-de-Henares, Spain	35	16	Nylon 6
SAFA	Blanes, Spain	44	20	Nylon 6/6
	San-Julian-de-Ramis, Spain	11	5	Nylon 6/6
SNIACE	Torrelavega, Spain	13	6	Nylon 6
Ems Chemie	Ems, Switzerland	29	13	Nylon 6
Viscosuisse	Emmenbrucke and Widnau, Switzerland	64	29	Nylon 6, 6/6
ICI	Doncaster, Gloucester and Pontypool, U.K.	209	95	Nylon 6/6
	TOTAL	1511	685	

Nylon 6 vs. Nylon 6/6

Nylon 6 and Nylon 6/6 fibers are essentially equivalent in their uses for fibers. Nylon 6 is generally less expensive to produce. Nylon 6/6 is produced primarily by those large companies that are highly integrated backward and in many cases actually spin the yarn.

Caprolactam and Nylon 6 technology have been extensively licensed worldwide, and virtually all nylon produced in the Far East and Eastern Europe is Nylon 6. In addition, most of the nylon production in Latin America is Nylon

6. Nylon 6 is claimed to have better dyeability, and Nylon 6/6 has better crimp rigidity and drawability in the production of fine denier yarns.

Economics

Typical production costs are presented in the following tables for Nylon 6 from caprolactam and for Nylon 6/6 from adipic acid.

Nylon 6

Capacity: 33 mm lb/yr (15 m MT/yr)
Capital cost: BLCC, $11 mm; OSBL, $4 mm; WC, $12 mm

	¢/lb	$/MT	%
Raw materials[a]	95.3	2101	87
Utilities	2.3	51	2
Operating costs	2.2	49	2
Overhead costs	9.2	203	9
Cost of production	109.0	2204	100
Transfer price	122.6	2703	

[a] Caprolactam at 85¢/lb

Nylon 6/6

Capacity: 33 mm lb/yr (15 m MT/yr)
Capital cost: BLCC, $16 mm; OSBL, $6 mm; WC, $12 mm

	¢/lb	$/MT	%
Raw materials[a]	84.3	1859	78
Utilities	6.5	143	6
Operating costs	3.5	77	3
Overhead costs	13.5	298	13
Cost of production	107.8	2377	100
Transfer price	127.8	2818	

[a] Adipic acid at 57¢/lb; HMDA at 85¢/lb

BISPHENOL-A

Bisphenol-A is a phenol derivative used in the manufacture of epoxy and polycarbonate resins as well as, to some extent, in phenoxy and polysulfone resins. Its technical designation is 4,4-isopropylidenediphenol.

Technology

Bisphenol-A is the para-para condensation product of two molecules of phenol and one of acetone. The initial preparation involves the acid-catalyst condensation of phenol and acetone at ambient conditions and was first developed in the early part of the century. Although other routes are technically possible, the reaction of phenol and acetone is still the major commercial process.

The acid-catalyzed reaction proceeds by the formation of a carbonium ion produced by the reaction of the strong acid and the acetone. The addition of the carbonium ion to a phenol ring forms a second carbonium ion, which adds to the second phenol ring. The reactions follow:

$$CH_3\overset{\overset{O}{\|}}{C}CH_3 + H^+ \longrightarrow CH_3\overset{\overset{OH}{|}}{\underset{+}{C}}CH_3$$

$$CH_3\overset{\overset{OH}{|}}{\underset{+}{C}}CH_3 + C_6H_5OH \longrightarrow HO{-}C_6H_4{-}C(CH_3)_2OH + H^+$$

$$HO{-}C_6H_4{-}C(CH_3)_2OH + H^+ \longrightarrow HO{-}C_6H_4{-}C(CH_3)_2OH_2^+$$

$$HO{-}C_6H_4{-}C(CH_3)_2OH_2^+ \longrightarrow HO{-}C_6H_4{-}\overset{+}{C}(CH_3)_2$$

$$HO{-}C_6H_4{-}\overset{+}{C}(CH_3)_2 + C_6H_5OH \longrightarrow HO{-}C_6H_4{-}\underset{\underset{CH_3}{|}}{\overset{\overset{CH_3}{|}}{C}}{-}C_6H_4{-}OH + H^+$$

At a temperature of about 50°C, crude bisphenol-A crystallizes out after 2 hours. With recycling of by-product isomers, yields are generally in excess of 90 percent. In the United States concentrated or gaseous HCl is generally used,

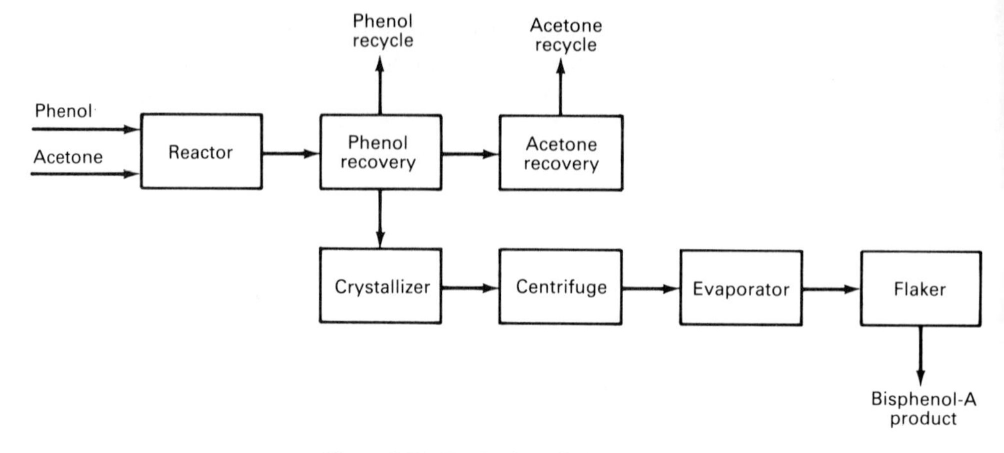

Figure 5.14 Production of bisphenol-A

but sulfuric acid is more common in other parts of the world. In some processes strongly acidic ion exchange resins are being used. The product used for polycarbonate normally must have considerably higher purity than that used for epoxy resins, and methods of processing downstream of the reactor to remove impurities can therefore vary. The purified product is usually obtained as an overhead product in a vacuum distillation and, while still molten, is mixed under pressure with a solvent and cooled to effect crystallization. The purified crystals are generally recovered in a centrifuge and dried. A block flow diagram of a typical process is shown in Figure 5-14.

Market

Over 90 percent of bisphenol-A is used, with almost even distribution, in the production of epoxy resins and polycarbonate resins. The remainder is utilized in the production of unsaturated polyester resins, polysulfone resins, flame retardants, and rubber chemicals. Increasing amounts of polycarbonate-based composites are expected to be used in transportation and packaging, and this ap-

TABLE 5-38 Major U.S. Producers of Bisphenol-A

Producer	Location	Estimated Capacity	
		mm lb/yr	m MT/yr
Dow Chemical	Freeport, Texas	170	77
General Electric	Mt. Vernon, Indiana	220	100
Shell Chemical	Deer Park, Texas	350	159
USS Chemicals	Haverhill, Ohio	170	77
	TOTAL	910	413

TABLE 5-39 Major Western European Producers of Bisphenol-A

Producer	Location	Estimated Capacity	
		mm lb/yr	m MT/yr
Progil Electrochimie	Le Pont-de-Claix, France	99	45
Bayer	Leverkusen, W. Germany	132	60
Dow Chemical	Stade, W. Germany	99	45
General Electric	Bergen-Op-Zoom, Netherlands	66	30
Shell Chemie	Rotterdam, Netherlands	132	60
Shell Chemicals	Ellesmere Port, U.K.	49	22
	TOTAL	577	262

plication will be the most rapidly increasing outlet for the material. Demand in the United States was about 643 million pounds in 1983 and 694 million pounds in 1984 and is projected to approach a billion pounds by 1988 (1,6).

Four current major producers of bisphenol-A are in the United States, as shown in Table 5-38 (1,2). In addition to these four producers, Union Carbide has a 70 million pound per year facility at Penuelas, Puerto Rico, on standby. Major Western European producers are shown in Table 5-39 (3). Production in Japan is approximately 100,000 metric tons per year, and the two major producers are Mitsui Toatsu Chemicals and Honshu Kagaku.

Economics

Typical production costs of bisphenol-A are as follows:

Capacity: 150 mm lb/yr (68 m MT/yr)
Capital cost: BLCC, $53 mm; OSBL, $25 mm; WC, $26 mm

	¢/lb	$/MT	%
Raw materials[a]	38.5	849	75
Utilities	1.5	33	3
Operating costs	2.2	49	4
Overhead costs	8.9	196	18
Cost of production	51.1	1127	100
Transfer price	66.7	1471	

[a] Phenol at 36¢/lb; acetone at 24¢/lb

POLYCARBONATES

Polycarbonates are specialty-engineering thermoplastics and are increasingly used as substitutes for competing materials such as die-cast zinc, magnesium, aluminum, and copper. Their unique combination of high-impact strength and excellent clarity are advantageous for glazing and lighting applications.

Technology

Polycarbonates are linear polyesters of carbonic acid. The general formula is as follows:

$$\mathrm{H(ORO\overset{\overset{\displaystyle O}{\|}}{C})_{\mathit{n}}OH}$$

The type of R group determines whether the polycarbonate is aliphatic, aromatic, or a combination of the two. The aromatic polycarbonates produced by the reaction of phosgene and bisphenol-A are of major importance and are represented by the following general formula:

$$\mathrm{H(O{-}C_6H_4{-}\overset{\overset{\displaystyle R}{|}}{\underset{\underset{\displaystyle R}{|}}{C}}{-}C_6H_4{-}O\overset{\overset{\displaystyle O}{\|}}{C})_{\mathit{n}}OH}$$

where R is a methyl group.

Commercial production of polycarbonates was started in 1958 by Bayer in West Germany. In 1960 Mobay in West Virginia began the production of polycarbonates based on bisphenol-A. In the same year General Electric, using its own technology, began producing the same material at facilities in Mt. Vernon, Indiana.

Much of the research in polycarbonates is based on the use of different dihydroxy compounds, the aim being modification of the properties to make them more suitable for special applications. The most important dihydroxy compound is bisphenol-A, and the aromatic polycarbonates are generally prepared by the phosgenation of the bisphenol-A by interfacial polymerization or by transesterification.

In interfacial polymerization aqueous solutions of alkali salts of bisphenol-A are prepared and then phosgenated in the presence of an inert organic solvent such as methylene chloride. The resulting low-molecular-weight product is then converted to a high-molecular-weight polycarbonate by polycondensation. The latter stages of the polycondensation are slow and are usually accelerated by adding small amounts of tertiary amines. The reaction can proceed at a relatively low temperature in an aqueous system, and drying of the reactants is not required. The reaction is also insensitive to many impurities. Disadvantages of this technology include the difficulty of removing electrolytes from the polymers and the necessity of isolating the polymers from dilute solutions.

Another approach to the reaction of bisphenol-A and phosgene involves the use of a solvent, or mixture of solvents, that can dissolve the reactants and also bind the by-product hydrogen chloride. One suitable solvent is pyridine. Water must be absent, and solubility of the resulting polycarbonate must be high enough to permit the formation of high-molecular-weight polycarbonates. A large excess of pyridine is generally required. The advantage of this technique is rapid reaction rate at low temperatures. However, the use of pyridine requires its recovery in anhydrous form before it can be recycled. A block flow diagram for this process is shown in Figure 5-15.

The transesterification technique can be used for aromatic polycarbonates as long as their melting point is lower than their decomposition temperatures. This technique is less desirable for high-molecular-weight polycarbonates. At temperatures low enough to avoid decomposition of the dialkyl carbonate and polycarbonate, the transesterification reaction is quite slow. The process employs equimolar quantities of bisphenol-A and diphenyl carbonate. The diphenyl carbonate is prepared by the phosgenation of phenol in a manner similar to the interfacial polycondensation of bisphenol-A. Phenol recovered from the transesterification step is recycled for the preparation of diphenyl carbonate. The polycarbonate produced by this technique is usually discolored and insoluble if large quantities are used. This result is due to the thermal instability of bisphenol-A and side reactions leading to branching and cross-linking at high temperatures. To minimize this problem, the reaction is generally carried out with a small excess of diphenyl carbonate and as short a reaction time as possible. The advantage of this technique over the interfacial polycondensation method is that the product is obtained in undiluted form and can be pelletized directly. Disadvantages are the need for high temperatures and a vacuum and limitations on molecular weight because of the high viscosity of the melt.

The technologies described are based on the use of phosgene, which is a hazardous and relatively expensive material. Therefore efforts are under way for a process not involving the use of phosgene. These efforts center around the following three possible reactions:

1. Transesterification not requiring the use of a phosgene intermediate.
2. Direct esterification of carbonic acid and aromatic diols.
3. Carbonylation of aromatic diols.

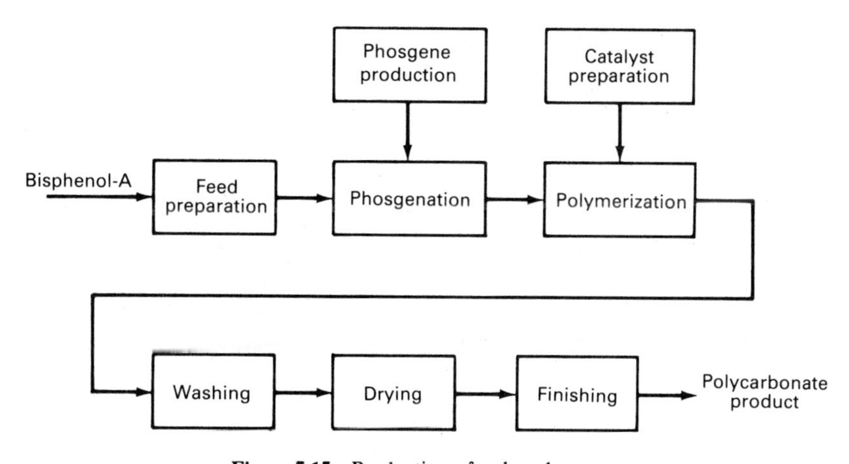

Figure 5.15 Production of polycarbonates

Market

Polycarbonates are important for end use applications requiring high performance. Production of these materials are limited at this time to the United States, Western Europe, and Japan. In the United States the only major polycarbonate producers are General Electric and Mobay Chemical, with General Electric supplying about two-thirds of the market. Dow Chemical will become the third producer when it begins operation of its facility in 1985.

Locations and estimated capacities of polycarbonate facilities are shown in Table 5-40. Dow is expected to add additional capacity by 1987. Mobay, in 1982, closed down its 30 million pound per year facility at New Martinsville, West Virginia, and General Electric postponed a 100 million pound per year expansion in Alabama. U.S. consumption in 1982 was about 260 million pounds and is projected to reach about 370 million pounds by 1987 (3). When consumption is compared with estimated capacity, it is apparent that an oversupply currently exists. Note that General Electric and Dow each produce both bisphenol-A and phosgene, the intermediates for polycarbonates, whereas Mobay makes only phosgene.

Much of the growing demand for polycarbonates is expected to come from the need for stronger, lighter-weight materials in automobiles, particularly for use in plastic bumpers, which are expected to grow substantially during the second half of this decade. The percentages of the market for the various end uses of polycarbonates are shown in the following table. The export market is approximately 50 million pounds per year, primarily to Japan, Brazil, Canada, Australia, and Mexico.

End Uses	%
Glazing	30
Communications and electronics	22
Appliances	10
Sports and recreation	8
Transportation	10
Lighting	3
Signs	2
Other	15

TABLE 5-40 U.S. Producers of Polycarbonates

		Estimated Capacity	
Producer	Location	mm lb/yr	m MT/yr
General Electric	Mount Vernon, Indiana	300	136
Mobay Chemical	Cedar Bayou, Texas	160	73
Dow Chemical	Freeport, Texas	30 (in 1985)	14
	TOTAL	490	223

TABLE 5-41 Major Japanese Producers of Polycarbonates

Producer	Location	Estimated Capacity mm lb/yr	m MT/yr
Mitsubishi Chemical	Kawasaki	13	6
Mitsubishi Gas Chemical	Osaka, Yokkaichi	33	15
Teijin	Matsuyama, Mikaua	33	15
	TOTAL	79	36

Principal producers of polycarbonates in Western Europe are Bayer and General Electric. Bayer produces 132 million pounds per year (60,000 MT/yr) at Leverkusen, West Germany, and General Electric produces 55 million pounds per year (25,000 MT/yr) at Bergen-Op-Zoom, The Netherlands. Terni Industrie Chimische also produces 20 million pounds per year (9000 MT/yr) at Nera-Montoro, Italy. They are expected to double their capacity in 1985. Principal producers in Japan are shown in Table 5-41.

Economics

Typical production costs for an interfacial polymerization process are as follows:

Capacity: 100 mm lb/yr (45 m MT/yr)
Capital cost: BLCC, $50 mm; OSBL, $20 mm; WC, $33 mm

	¢/lb	$/MT	%
Raw materials[a]	77.6	1711	79
Utilities	6.1	135	6
Operating costs	2.6	57	3
Overhead costs	11.8	260	12
Cost of production	98.1	2163	100
Transfer price	109.1	2406	

[a] Bisphenol-A at 67¢/lb

METHYL METHACRYLATE

Methyl methacrylate is the most important of the esters of methacrylic acid. It is polymerized or copolymerized to produce acrylic resins, which have high strength and excellent transparency and weather resistance. At present the major process is the acetone cyanohydrin route, which employs acetone, hydrogen cyanide, and methanol as raw materials. However, substantial work is proceeding on processes starting with C_4's. The Japanese appear to be leading the way in these developments.

Technology

The acetone cyanohydrin process was originally introduced by ICI about 50 years ago. The process involves the addition of HCN to acetone, followed by hydrolysis and esterification. The hydrolysis step generally uses a sulfuric acid catalyst, and methanol is required for the esterification step. The reactions are as follows:

$$CH_3\overset{\overset{O}{\|}}{C}CH_3 + HCN \longrightarrow CH_3\underset{\underset{CN}{|}}{\overset{\overset{OH}{|}}{C}}CH_3$$

Acetone cyanohydrin

$$CH_3\underset{\underset{CN}{|}}{\overset{\overset{OH}{|}}{C}}CH_3 \xrightarrow{H_2SO_4,\ CH_3OH} CH_2{=}CCOOCH_3 + NH_4HSO_4$$

The original process has been modified and improved by several companies, including Rohm and Haas and DuPont in the United States. The addition of HCN to acetone generally proceeds at a temperature of about 40°C and is base

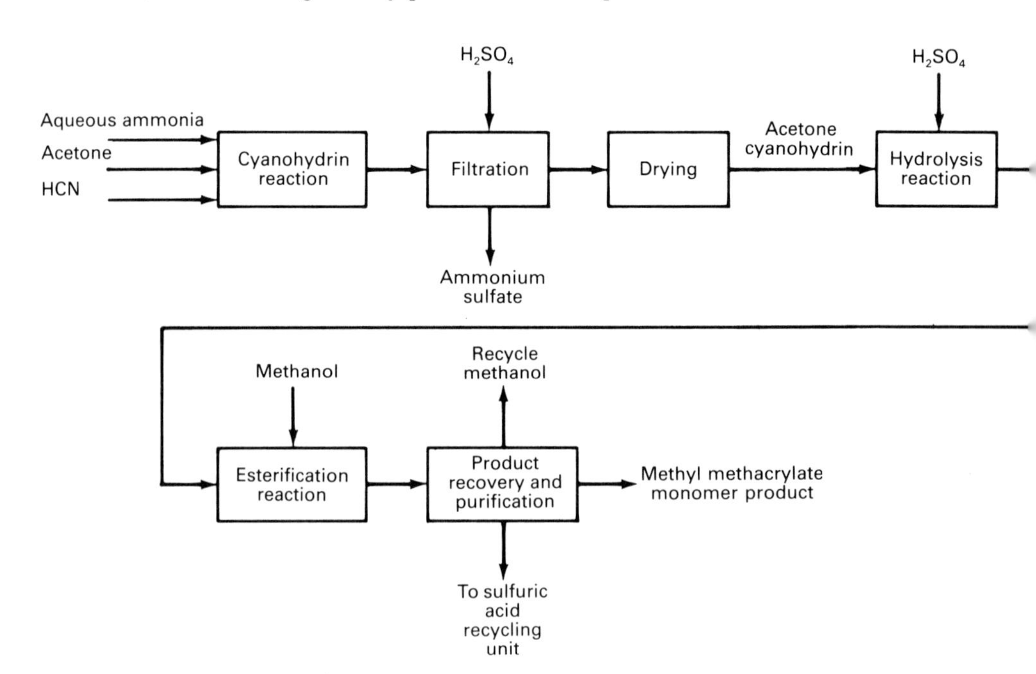

Figure 5.16 Production of methyl methacrylate—acetone cyanohydrin process

catalyzed. Yields are in excess of 98 percent. The acetone cyanohydrin is dehydrated with concentrated sulfuric acid to produce methacrylamide sulfate. At temperatures of about 100°C and residence times of about an hour, yields are in excess of 90 percent. Methyl methacrylate is then obtained in almost quantitative yield with methanol and some excess sulfuric acid. Temperature is about 110°C, pressure about 5 atmospheres, and residence time 1 to 2 hours. Improvements in the technology are very closely guarded by the producing companies, and details are not generally available. A typical block flowsheet of the process is shown in Figure 5-16.

Alternative processes being developed, particularly in Japan, are primarily based on C_4 feedstocks (2). These technologies would usually substitute a less expensive feedstock and would eliminate the problem of disposing of by-product ammonium sulfates. These processes generally involve an oxidation reaction. The reaction, starting with isobutylene, is as follows:

$$CH_2{=}C(CH_3)CH_3 \xrightarrow{O_2,\ \text{catalyst}} CH_2{=}C(CH_3)CHO \xrightarrow{O_2,\ \text{catalyst}} \underset{\text{Methacrylic acid}}{CH_2{=}C(CH_3)COOH}$$

$$CH_2{=}C(CH_3)COOH + CH_3OH \longrightarrow CH_2{=}C(CH_3)COOCH_3 + H_2O$$

A technology developed by Japan Catalytic Chemical is a high-yield oxidation of isobutylene or tertiary butyl alcohol. The first oxidation step converts the isobutylene to methyl acrolein, which is then further oxidized to methacrylic acid. Esterification is then required to produce methyl methacrylate. This process is similar to the Oxirane process, which was planned in the early eighties using tertiary butyl alcohol as feedstock. Tertiary butyl alcohol is a coproduct of propylene oxide production at the site at Baytown, Texas. The project was ultimately canceled for the ostensible reason that the tertiary butyl alcohol was more attractive as an unleaded gasoline octane enhancer than as a feedstock. Processes using isobutylene as feedstock have also been developed by Asahi Chemical and Mitsubishi Rayon.

Market

The major end use for polymethyl methacrylate in the United States includes cast and extruded acrylic sheet (30%), surface coating resins (33%), and molding resins (25%). The monomer is also used as a comonomer with acrylonitrile in the production of acrylic fibers. Major producers of methyl methacrylate in the United States are shown in Table 5-42 (8). Demand in 1984 was 825 million pounds and is projected to reach 960 million pounds in 1989.

Major producers in Western Europe are shown in Table 5-43 (5). Major producers in Japan are shown in Table 5-44.

TABLE 5-42 Major U.S. Producers of Methyl Methacrylate

Producer	Location	Estimated Capacity	
		mm lb/yr	m MT/yr
Rohm and Haas	Deer Park, Texas	660	299
DuPont	Memphis, Tennessee	250	113
CY/RO[a]	Westwego, Louisiana	200	91
	TOTAL	1110	503

[a]CY/RO is a joint venture of American Cyanamid and Roehm of West Germany.

The industry in the United States is characterized by large, integrated operations. DuPont and Rohm and Haas produce HCN and methanol, and CY/RO produces HCN. Integration downstream is extensive. CY/RO produces extruded and cast sheets and molding and extrusion resins, which account for almost two-thirds of the end uses. Trade names for these products are Plexiglas for Rohm and Haas, Lucite for DuPont, and Acrylite and Exolite for CY/RO.

Substitution of new technology is primarily dependent on feedstock developments. At present acetone is readily available as a coproduct of phenol production and is dependent primarily on the housing market. Hydrogen cyanide is a by-product of the production of acrylonitrile, which is largely dependent on the fiber market. A drop in the growth of acrylic fibers and the development of new technology may reduce the production of by-product hydrogen cyanide.

The feedstock situation in Western Europe is aggravated by higher raw material costs, and a change to the potentially less expensive isobutylene route is held back because of the demand for isobutylene to make gasoline octane enhancers. The feedstock situation in Japan, however, is quite different. The economic situation there for the acetone cyanohydrin route is much less attractive. Some acrylonitrile facilities have been shut down, and increased use of newer technology produces less HCN. In addition, acetone is not very plentiful. Furthermore the use of isobutylene for octane enhancers is not as prevalent in Japan, and thus this application does not compete extensively with methacrylate

TABLE 5-43 Major Western European Producers of Methyl Methacrylate

Producer	Location	Estimated Capacity	
		mm lb/yr	m MT/yr
Norsolor	Saint-Avold, France	110	50
Degussa	Wesseling, W. Germany	99	45
Anic	Pisticci, Italy	13	6
Vedril	Rho, Italy	95	43
Monacril	Palos-de-la-Frontera, Spain	44	20
Paular	Morell, Spain	66	30
ICI	Billingham, U.K.	232	105
	TOTAL	659	299

TABLE 5-44 Major Japanese Producers of Methyl Methacrylate

Producer	Estimated Capacity		Process
	mm lb/yr	m MT/yr	
Asahi Chemical	100	45	Acetone cyanohydrin
Kyowa Gas Chemical	90	41	Acetone cyanohydrin
Mitsubishi Rayon	165	75	Acetone cyanohydrin
	70	32	TBA oxidation
Sumitomo Chemical	90	41	C_4 oxidation
TOTAL	515	234	

for isobutylene. Therefore Japan has a considerable incentive to develop the new technology.

Economics

Typical production costs are presented in the following tables for the acetone cyanohydrin process and the isobutylene oxidation route.

Acetone Cyanohydrin

Capacity: 300 mm lb/yr (136 m MT/yr)
Capital cost: BLCC, $103 mm; OSBL, $49 mm; WC, $47 mm

	¢/lb	$/MT	%
Raw materials[a]	30.2	666	65
Utilities	5.9	130	13
Operating costs	2.9	64	6
Overhead costs	7.6	168	16
Cost of production	46.6	1028	100
Transfer price	61.8	1363	

[a] Acetone at 24¢/lb; HCN at 40¢/lb; methanol at 54¢/gal; sulfuric acid at 4¢/lb

Isobutylene Oxidation

Capacity: 300 mm lb/yr (136 m MT/yr)
Capital cost: BLCC, $108 mm; OSBL, $43 mm; WC, $58 mm

	¢/lb	$/MT	%
Raw materials[a]	32.5	717	56
Utilities	14.4	318	25
Operating costs	2.0	44	3
Overhead costs	8.9	196	16
Cost of production	57.8	1275	100
Transfer price	72.9	1607	

[a] Isobutylene at 32¢/lb; methanol at 54¢/gal

REFERENCES

Ethylbenzene and Styrene

1. List, H. L. "Ethylbenzene and Styrene—A Status Report," *International Petrochemical Developments,* Vol. 2, No. 9, May 1, 1981.
2. "Chemical Profiles," *Chemical Marketing Reporter,* September 5, 1983.
3. Ibid., September 12, 1983.
4. *1984 Directory of Chemical Producers—United States,* SRI International.
5. "World Styrene Consumption Seen Growing 5.1% Annually," *Hydrocarbon Processing,* February 1982, p. 23.
6. "Styrene Export Wings Get Clipped," *Chemical Business,* August 23, 1982, pp. 40–42.
7. "Canada Sees Lucrative Styrene Export Market," *Chemical and Engineering News,* October 18, 1982, pp. 15–16.
8. "A Better Styrene Goes Commercial," *Chemical Week,* February 17, 1982, pp. 42–46.
9. Innes, R. A. and H. E. Swift. "Toluene to Styrene—A Difficult Goal," *Chemtech,* April 1981, pp. 244–48.
10. Gale, G. "Energizing the Styrene Business," *Chemical Business,* December 15, 1980, pp. 9–18.
11. "Petrochemical Handbook '83," *Hydrocarbon Processing,* November 1983.
12. U.S. Patent 4,115,424 (September 19, 1978) to Monsanto.
13. *1984 Directory of Chemical Producers—Western Europe,* SRI International.
14. *Kirk-Oihmer Encyclopedia of Chemical Technology,* Third Edition, Vol. 21. New York: Wiley, 1984, pp. 770–801.
15. Ibid., Vol. 24, pp. 709–44.
16. *JCW Chemicals Guide 82/83.* Japan: The Chemical Daily Co. Ltd., March 1982.
17. *Chemical Industry Yearbook,* Second Edition. Surrey, England: Industrial Press, 1984.
18. "Styrene Monomer Producers Slate An April 1 Price Increase," *Chemical Marketing Reporter,* March 4, 1985.

Polystyrene

1. List, H. L. "Polystyrene—A Status Report," *International Petrochemical Developments,* Vol. 3, No. 6, March 15, 1982.
2. "Polystyrene," *Chemical and Engineering News,* April 18, 1983, p. 17.
3. "Materials 84," *Modern Plastics,* January 1984.
4. "Petrochemical Handbook '83," *Hydrocarbon Processing,* November 1983.
5. *1984 Directory of Chemical Producers—Western Europe,* SRI International.
6. *1984 Directory of Chemical Producers—United States,* SRI International.
7. Kirk-Othmer Encyclopedia of Chemical Technology, Third Edition, Vol. 21. New York: Wiley, 1984, pp. 801–47.
8. *JCW Chemicals Guide.* Japan: The Chemical Daily Co. Ltd., March 1982.
9. *Chemical Industry Yearbook,* Second Edition. Surrey, England: Industrial Press, 1984.

ABS

1. List, H. L. "ABS Resins—A Status Report," *International Petrochemical Developments,* Vol. 2, No. 18, September 15, 1981.
2. "Chemical Profiles," *Chemical Marketing Reporter,* November 29, 1982.
3. "The Shakeout Over, ABS Makers Stage a Rebound," *Chemical Week,* August 17, 1983.
4. "ABS Losing Pipe Market," *Chemical Marketing Reporter,* November 8, 1982.
5. "Round Two of the Shakeout in Bulk Resins," *Chemical Week,* October 5, 1983.
6. "Petroleum Handbook '83," *Hydrocarbon Processing,* November 1983.
7. *1984 Directory of Chemical Producers—Western Europe,* SRI International.
8. *Kirk-Othmer Encyclopedia of Chemical Technology,* Third Edition, Vol. 1. New York: Wiley, 1984, pp. 442–55.
9. *1984 Directory of Chemical Producers—United States,* SRI International.
10. *JCW Chemicals Guide 82/83.* Japan: The Chemical Daily Co. Ltd., March 1982.
11. "Materials '85," *Modern Plastics,* January 1985, pp. 61–71.

SBR

1. *1984 Directory of Chemical Producers—United States,* SRI International.
2. *1984 Directory of Chemical Producers—Western Europe,* SRI International.
3. *Kirk-Othmer Encyclopedia of Chemical Technology,* Third Edition, Vol. 8. New York: Wiley, 1984, pp. 608–25.
4. "Rubber Report," *Chemical and Engineering News,* April 25, 1983, pp. 23–40.
5. Greek, B. F. "Elastomers Finally Recover Growth," *Chemical and Engineering News,* April 30, 1984, pp. 35–56.
6. "Synthetic Rubbers Ride on Auto, Housing Success," *Chemical Week,* October 17, 1984, pp. 54–58.
7. "IISRP Forecasts Flat Period for Synthetic Rubber Makers," *European Chemical News,* February 21, 1983, p. 9.
8. *JCW Chemicals Guide 82/83.* Japan: The Chemical Daily Co. Ltd., March 1982.
9. "Materials '85," *Modern Plastics,* January 1985, pp. 61–71.
10. *Chemical Industry Yearbook,* Second Edition. Surrey, England: Industrial Press, 1984.

Nitrobenzene

1. List, H. L. "Nitrobenzene—A Status Report," *International Petrochemical Developments,* Vol. 3, No. 1, January 1, 1982.
2. "Chemical Profiles," *Chemical Marketing Reporter,* July 30, 1984.
3. *1984 Directory of Chemical Producers—United States,* SRI International.
4. *1984 Directory of Chemical Producers—Western Europe,* SRI International.
5. *Kirk-Othmer Encyclopedia of Chemical Technology,* Third Edition, Vol. 15. New York: Wiley, 1984, pp. 916–32.
6. *JCW Chemicals Guide 82/83.* Japan: The Chemical Daily Co. Ltd., March 1982.

Aniline

1. "Chemical Profiles," *Chemical Marketing Reporter,* July 9, 1984.
2. *1984 Directory of Chemical Producers—United States,* SRI International.
3. "USS Chemicals Makes Its Debut in Aniline," *Chemical Week,* June 9, 1982, pp. 28–30.
4. List, H. L. "Aniline—A Status Report," *International Petrochemical Developments,* Vol. 3, No. 2, January 15, 1982.
5. "Petrochemical Handbook '83," *Hydrocarbon Processing,* November 1983.
6. *1984 Directory of Chemical Producers—Western Europe,* SRI International.
7. *Kirk-Othmer Encyclopedia of Chemical Technology,* Third Edition, Vol. 2. New York: Wiley, 1984, pp. 309–20.
8. *JCW Chemicals Guide 82/83.* Japan: The Chemical Daily Co. Ltd., March 1982.
9. "Aniline Output Sets Record and New Growth Is Seen in '85," *Chemical Marketing Reporter,* February 18, 1985.

Polyisocyanates

1. List, H. L. "Isocyanates," *International Petrochemical Developments,* Vol. 2, No. 1, January 1, 1981.
2. *1984 Directory of Chemical Producers—United States,* SRI International.
3. *1984 Directory of Chemical Producers—Western Europe,* SRI International.
4. "Chemical Profile," *Chemical Marketing Reporter,* November 26, 1984.
5. *Kirk-Othmer Encyclopedia of Chemical Technology,* Third Edition, Vol. 13. New York: Wiley, 1984, pp. 789–818.
6. *JCW Chemicals Guide 82/83.* Japan: The Chemical Daily Co. Ltd., March 1982.

Cyclohexane

1. List, H. L. "Cyclohexane—A Status Report," *International Petrochemical Developments,* Vol. 3, No. 9, May 1, 1982.
2. *1984 Directory of Chemical Producers—United States,* SRI International.
3. "Chemical Profile," *Chemical Marketing Reporter,* December 5, 1983.
4. *1984 Directory of Chemical Producers—Western Europe,* SRI International.
5. "Petrochemical Handbook '83," *Hydrocarbon Processing,* November 1983.
6. *Kirk-Othmer Encyclopedia of Chemical Technology,* Third Edition, Vol. 12. New York: Wiley, 1984, pp. 931–37.
7. *JCW Chemicals Guide 82/83.* Japan: The Chemical Daily Co. Ltd., March 1982.
8. "Key Chemicals—Cyclohexane," *Chemical and Engineering News,* December 17, 1984, p. 16.
9. *Chemical Industry Yearbook,* Second Edition. Surrey, England: Industrial Press, 1984.

Adipic Acid

1. List, H. L. "Adipic Acid—A Status Report," *International Petrochemical Developments,* Vol. 2, No. 11, June 1, 1981.
2. *1984 Directory of Chemical Producers—United States,* SRI International.

3. "Chemical Profile," *Chemical Marketing Reporter,* October 17, 1983.
4. W. German Offen Patent 2,630,086 (January 12, 1978) to BASF.
5. W. German Offen Patent 2,646,955 (April 20, 1978) to BASF.
6. *1984 Directory of Chemical Producers—Western Europe,* SRI International.
7. *Kirk-Othmer Encyclopedia of Chemical Technology,* Third Edition, Vol. 1. New York: Wiley, 1984, pp. 510–31.
8. *JCW Chemicals Guide 82/83.* Japan: The Chemical Daily Co. Ltd., March 1982.
9. *Chemical Industry Yearbook,* Second Edition. Surrey, England: Industrial Press, 1984.

Cumene, Phenol, and Acetone

1. Gelbein, A. P. and A. S. Nislick. "Make Phenol From Benzoic Acid," *Hydrocarbon Processing,* November 1978, pp. 125–28.
2. *Chemical and Engineering News,* December 19, 1983, p. 9.
3. List, H. L. "Phenol—Update," *International Petrochemical Developments,* Vol. 4, No. 1, January 1, 1983.
4. "Chemical Profile," *Chemical Marketing Reporter,* July 2, 1984.
5. *Chemical Marketing Reporter,* April 2, 1984, p. 3.
6. Greek, B. F. "Phenol, Vinyl Acetate Head for Moderate Pickup in 1984," *Chemical and Engineering News,* December 19, 1983, pp. 7–9.
7. " A Shakeout As Phenol Scrapes Bottom," *Chemical Week,* September 22, 1982, pp. 28–30.
8. List, H. L. "Phenol—A Status Report," *International Petrochemical Developments,* Vol. 2, No. 3, February 1, 1981.
9. *1984 Directory of Chemical Producers—United States,* SRI International.
10. "Petrochemical Handbook '83," *Hydrocarbon Processing,* November 1983.
11. *1984 Directory of Chemical Producers—Western Europe,* SRI International.
12. "Phenol, Vinyl Acetate Face Tightening Supply," *Chemical and Engineering News,* September 24, 1984, pp. 10–11.
13. "Chemical Profile," *Chemical Marketing Reporter,* August 27, 1984.
14. *Kirk-Othmer Encyclopedia of Chemical Technology,* Third Edition, Vol. 1. New York: Wiley, 1984, pp. 179–91.
15. Ibid., Vol. 7, pp. 286–90.
16. Ibid., Vol. 17, pp. 373–84.
17. *JCW Chemicals Guide 82/83.* Japan: The Chemical Daily Co. Ltd., March 1982.
18. *Chemical Industry Yearbook,* Second Edition. Surrey, England: Industrial Press, 1984.

Caprolactam

1. List, H. L. "Caprolactam—A Status Report," *International Petrochemical Developments,* Vol. 3, No. 10, May 15, 1982.
2. "Chemical Profiles," *Chemical Marketing Reporter,* October 24, 1983.
3. *1984 Directory of Chemical Producers—United States,* SRI International.
4. "Caprolactam Close to 100% As Nylon 6 Booms," *Chemical Marketing Reporter,* November 28, 1983.
5. "Petrochemical Handbook '83," *Hydrocarbon Processing,* November 1983.
6. *1984 Directory of Chemical Producers—Western Europe,* SRI International.

Nylon Fibers

1. *1984 Directory of Chemical Producers—United States,* SRI International.
2. *1984 Directory of Chemical Producers—Western Europe,* SRI International.
3. "Petrochemical Handbook '83," *Hydrocarbon Processing,* November 1983.
4. *JCW Chemicals Guide 82/83.* Japan: The Chemical Daily Co. Ltd., March 1982.
5. List, H. L. "Nylon 6 Versus Nylon 6/6," *International Petrochemical Developments,* December 1, 1981.
6. "Outlook Unsettled After Boom in Fibers," *European Chemical News,* February 18, 1985, p. 10.

Bisphenol-A

1. "Chemical Profile," *Chemical Marketing Reporter,* July 16, 1984.
2. *1984 Directory of Chemical Producers—United States,* SRI International.
3. *1984 Directory of Chemical Producers—Western Europe,* SRI International.
4. *Kirk-Othmer Encyclopedia of Chemical Technology,* Third Edition, Vol. 2. New York: Wiley, 1984, pp. 75–95.
5. List, H. L. "Bisphenol-A, A Status Report," *International Petrochemical Developments,* Vol. 6, No. 2, January 15, 1985.
6. "Bisphenol-A Producers Are Upbeat About 1985," *Chemical Marketing Reporter,* December 31, 1984.

Polycarbonates

1. List, H. L. "Polycarbonates—A Status Report," *International Petrochemical Developments,* Vol. 2, No. 16, August 15, 1981.
2. "Chemical Profiles," *Chemical Marketing Reporter,* May 14, 1984.
3. "Dow: Three in Polycarbonates Is Not a Crowd," *Chemical Week,* November 10, 1982, pp. 30–31.
4. "Polycarbonate Resins Boosted by Specialties," *Chemical Marketing Reporter,* February 13, 1984.
5. *1984 Directory of Chemical Producers—Western Europe,* SRI International.
6. *1984 Directory of Chemical Producers—United States,* SRI International.
7. *Kirk-Othmer Encyclopedia of Chemical Technology,* Third Edition, Vol. 18. New York: Wiley, 1984, pp. 479–94.
8. *JCW Chemicals Guide 82/83.* Japan: The Chemical Daily Co. Ltd., March 1982.
9. "Materials '85," *Modern Plastics,* January 1985, pp. 61–71.

Methyl Methacrylate

1. List, H. L. "Methyl Methacrylate—A Status Report," *International Petrochemical Developments,* Vol. 2, No. 17, September 1, 1981.
2. Nakamura, T. and T. Kita. "A New Feedstock for the Manufacture of Methyl Methacrylate Emerges," *Chemical Economy and Engineering Review,* Vol. 15, No. 10 (No. 172), October 1983, pp. 23–27.

3. Itakura, J. "Present State and Prospects for Acrylic Ester Industry," *Chemical Economy and Engineering Review,* Vol. 13, Nos. 7–8 (No. 150), July/August 1981, pp. 19–24.
4. "Petrochemical Handbook '83," *Hydrocarbon Processing,* November 1983.
5. *1984 Directory of Chemical Producers—Western Europe,* SRI International.
6. *Kirk-Othmer Encyclopedia of Chemical Technology,* Third Edition, Vol. 15. New York: Wiley, 1984, pp. 377–97.
7. *JCW Chemicals Guide 82/83.* Japan: The Chemical Daily Co. Ltd., March 1982.
8. "Chemical Profile," *Chemical Marketing Reporter,* January 28, 1985.

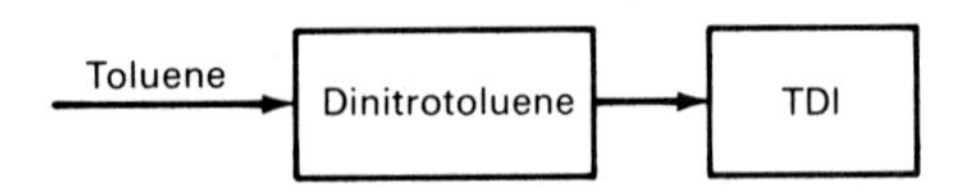

Toluene derivatives

CHAPTER SIX

Toluene Derivatives

DINITROTOLUENE AND TOLUENE DIISOCYANATE

Isocyanates have emerged as very important petrochemicals for use in polyurethanes. Toluene diisocyanate (TDI) is used primarily for the production of flexible foams, which are used for mattresses, cushions, and automobile interiors. The excellent properties of TDI in flexible polyurethane have established it as the material of choice in the majority of seat-cushioning applications.

Technology

At present most of the commercial isocyanates are produced by the phosgenation of amines. In the case of TDI, the starting point is dinitrotoluene (DNT), which is produced by the nitration of toluene. DNT is an article of commerce and is available on the market to isocyanate producers. Frequently, however, it is produced as a first step in the manufacture of TDI.

The second step in the process is the reduction of dinitrotoluene to tolylene diamine. The reduction is usually accomplished by the reaction of hydrogen and DNT in the presence of a suitable catalyst, such as Raney nickel or supported platinum or palladium in the presence of a solvent such as methanol. The reaction is as follows:

$$CH_3C_6H_3(NO_2)_2 + 6H_2 \xrightarrow{\text{catalyst}} CH_3C_6H_3(NH_2)_2 + 4H_2O$$

This reaction involves 2,4-DNT, although the dinitrotoluenes are actually a mixture of isomers. The 2,4-DNT and 2,6-DNT isomers predominate, with the ratio of 2,4-DNT to 2,6-DNT being about 4/1. The isomers of DNT are converted to the corresponding diamines in roughly the same relative amounts. The tolylene diamine may also be produced directly from toluene by amination. Although this technique would eliminate the need for toluene nitration and has a potential for some cost savings, it is not generally employed.

The tolylene diamine is conventionally reacted with phosgene to produce the diisocyanate. Although the phosgenation is a simple reaction, the use of the highly toxic phosgene requires care. Phosgene is produced by the gas-phase reaction of carbon monoxide and chlorine using an activated carbon bed. The phosgenation reaction proceeds in two steps. The first step produces carbamoyl chlorides, and the second step, at elevated temperatures, decomposes the carbamoyl chlorides to isocyanates, as shown in the following reactions:

$$CH_3C_6H_3(NH_2)_2 + 2COCl_2 \longrightarrow CH_3C_6H_3(NHCOCl)_2 + 2HCl$$

$$CH_3C_6H_3(NHCOCl)_2 \xrightarrow{160-180°C} CH_3C_6H_3(NCO)_2 + 2HCl$$

A block flow diagram for the process is shown in Figure 6-1.

The disadvantages of using phosgene have encouraged many companies to investigate routes to isocyanate that do not require the use of phosgene. Mitsui Toatsu of Japan has been active in the development of a process based on the carbonylation of dinitrotoluene using a palladium or selenium catalyst, but a commercial facility has not been built (7,8). This technology involves an intermediate diurethane product, which is then decomposed to the isocyanate, as shown in the following reactions:

$$CH_3C_6H_3(NO_2)_2 + 4CO + 2CH_3OH \longrightarrow CH_3C_6H_3(NHCOOCH_3)_2 + 2CO_2$$

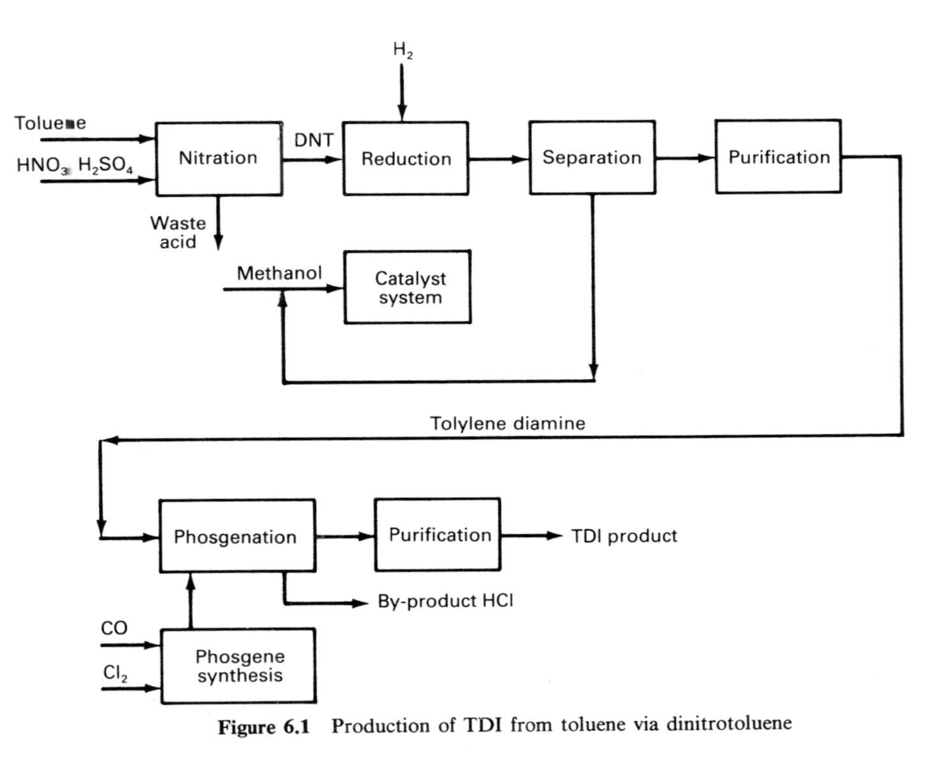

Figure 6.1 Production of TDI from toluene via dinitrotoluene

TABLE 6-1 Major U.S. Producers of TDI

Producer	Location		Estimated Capacity mm lb/yr	Estimated Capacity m MT/yr
BASF Wyandotte	Geismar, Louisiana		125	57
Dow	Freeport, Texas		100	45
Mobay	Cedar Bayou, Texas		125	57
	New Martinsville, W. Virginia		100	45
Olin	Lake Charles, Louisiana		125	57
	Moundsville, W. Virginia		80	36
Rubicon	Geismar, Louisiana		40	18
		TOTAL	695	315

TABLE 6-2 Major Western European Producers of TDI

Producer	Location		Estimated Capacity mm lb/yr	Estimated Capacity m MT/yr
Bayer-Shell Isocyanates	Antwerp, Belgium		66	30
PBU	Le Pont-de-Claix, France		106	48
Bayer	Brunsbuttel, W. Germany		99	45
	Dormagen, W. Germany		115	52
	Leverkusen, W. Germany		106	48
Montepolimeri	Porto-Marghera, Italy		132	60
Bayer Hispania Ind.	Tarragona, Spain		53	24
		TOTAL	677	307

TABLE 6-3 Major Japanese Producers of TDI

Producer	Location		Estimated Capacity mm lb/yr	Estimated Capacity m MT/yr
Nippon Polyurethane	Nanyo		33	15
Mitsui Toatsu Chemical	Omuta, Nagoya		55	25
Takeda Chemical	Tokuyama, Kashima		42	19
Mitsubishi Chemical	Kurosaki		66	30
Sumitomo Bayer Urethane	Kikumoto		25	11
		TOTAL	221	100

$$CH_3C_6H_3(NHCOOCH_3)_2 \xrightarrow{\Delta} CH_3C_6H_3(NCO)_2 + 2CH_3OH$$

The advantages of such a route are the elimination of the need for producing tolylene diamine and phosgene, with the accompanying savings in raw materials by eliminating the need for chlorine and hydrogen. Commercialization will undoubtedly depend on substantial increases in demand for TDI and is not generally expected to occur in the short term.

Market

In the early eighties the flexible-foam market was considerably reduced by the slowdown in home construction and in the automobile industry. However, growth continued in the mid-eighties because of the economic recovery in these industries, and modest growth is generally forecasted for the remainder of the decade as the major end use markets mature. Flexible urethane foams account for about 90 percent of the market, with polyurethane coatings and elastomers accounting for much of the remainder. U.S. demand for TDI in 1984 was about 640 million pounds, and projections for 1988 indicate only a slight increase over this level (2). Current installed capacity is almost 700 million pounds per year despite the withdrawal of Union Carbide and DuPont from the business in the late seventies. Therefore additional capacity is not expected to be needed for some years.

Current producers of TDI in the United States are shown in Table 6-1 (2,3). The Olin facility at Moundsville was purchased from Allied in 1981, and at the same time Olin closed an equal capacity facility at Ashtabula, Ohio. The Olin facility at Moundsville was shut down in 1984 for an indefinite period.

Major Western European producers of TDI are shown in Table 6-2 (6). Major Japanese producers of DNT are Daiwa Chemical, Mitusbishi Chemical Industries, Nippon Kayaku, and Sumitomo Chemical. Major Japanese producers of TDI are shown in Table 6-3.

Economics

Typical production costs for DNT and the complete process for TDI, including the intermediate production of DNT, are as follows:

Dinitrotoluene (excluding sulfuric acid concentration)

Capacity: 130 mm lb/yr (59 m MT/yr)
Capital cost: BLCC, $15 mm; OSBL, $6 mm; WC, $8 mm

	¢/lb	$/MT	%
Raw materials[a]	17.2	379	82
Utilities	0.3	7	1
Operating costs	1.1	24	5
Overhead costs	2.4	53	12
Cost of production	21.0	463	100
By-product credit (spent acid)	(1.7)	37	
Net cost of production	19.3	426	
Transfer price	24.2	534	

[a]Toluene at 13¢/lb; nitric acid at 10¢/lb; sulfuric acid at 4¢/lb

TDI (including production of DNT)

Capacity: 100 mm lb/yr (45 m MT/yr)
Capital cost: BLCC, $57 mm; OSBL, $23 mm; WC, $22 mm

	¢/lb	$/MT	%
Raw materials[a]	37.2	820	53
Utilities	16.6	366	24
Operating costs	3.6	79	5
Overhead costs	13.2	291	18
Cost of production	70.6	1557	100
By-product credit (HCl)	(5.9)	(130)	
Net cost of production	64.7	1427	
Transfer price	88.7	1956	

[a]Toluene at 13¢/lb; nitric acid at 10¢/lb; chlorine at 7¢/lb

REFERENCES

1. List, H. L. "Toluene Diisocyanate—Update," *International Petrochemical Developments*, Vol. 4, No. 7, April 1, 1983.
2. "Chemical Profiles," *Chemical Marketing Reporter*, November 19, 1984.
3. *1984 Directory of Chemical Producers—United States*, SRI International.
4. "Slump in Polyurethane Continues," *Chemical and Engineering News*, September 20, 1982, p. 27.
5. "Petrochemical Handbook '83," *Hydrocarbon Processing*, November 1983.
6. *1984 Directory of Chemical Producers—Western Europe*, SRI International.

7. Japanese Kokai Tokyo Koho 81-68,653 to Mitsui Toatsu Chemicals, June 9, 1981.
8. Japanese Kokai Tokyo Koho 81-138,159 to Mitsui Toatsu Chemicals, October 28, 1981.
9. *Kirk-Othmer Encyclopedia of Chemical Technology*, Third Edition, Vol. 13. New York: Wiley, 1984, pp. 789–818.
10. *JCW Chemicals Guide 82/83.* Japan: The Chemical Daily Co. Ltd., March 1982.

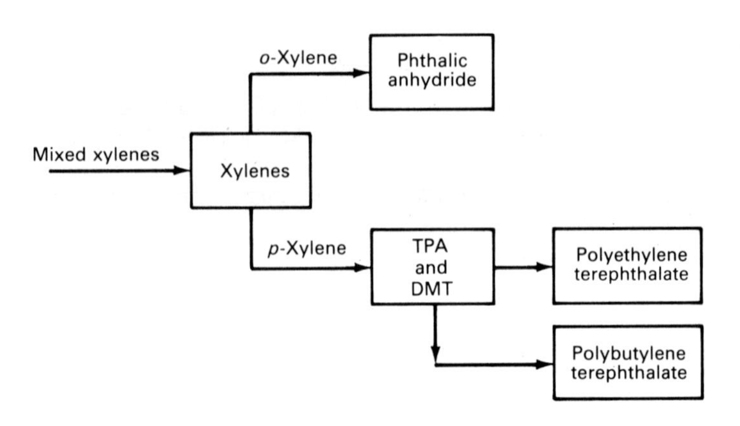

Xylenes and xylene derivatives

CHAPTER SEVEN

Xylene and Xylene Derivatives

XYLENES

Xylenes are obtained primarily as products of catalytic reforming and steam cracking of hydrocarbons. The mixed xylenes thus obtained are used mostly as a source of the individual xylene monomers—ortho, meta, and para. In addition, ethylbenzene, which has a boiling point in the same range as the xylenes, is also recovered from the mixed xylene stream. The mixed xylenes are also used as solvents and as gasoline blending components. When used as gasoline blending components, the mixed xylenes need not be separated from the reformate.

Technology

Paraxylene is separated from the mixed xylene stream by either crystallization or adsorption, and orthoxylene and ethylbenzene are recovered by fractionation. A typical xylene complex could also include an isomerization unit, which could convert metaxylene and even orthoxylene and ethylbenzene to paraxylene, which is generally the more desirable isomer. A block flow diagram for a typical installation is shown in Figure 7-1.

The aromatics contained in high-octane reformates are in concentrations ranging from 45 to 70 percent by volume. A typical aromatics distribution of a 94-octane reformate produced from a 200 to 375°F boiling range naphtha is shown in the following table:

Component	Volume % of Reformate
Benzene	5
Toluene	24
Ethylbenzene	4
Paraxylene	4
Metaxylene	9
Orthoxylene	5
C_9-C_{10} aromatics	4

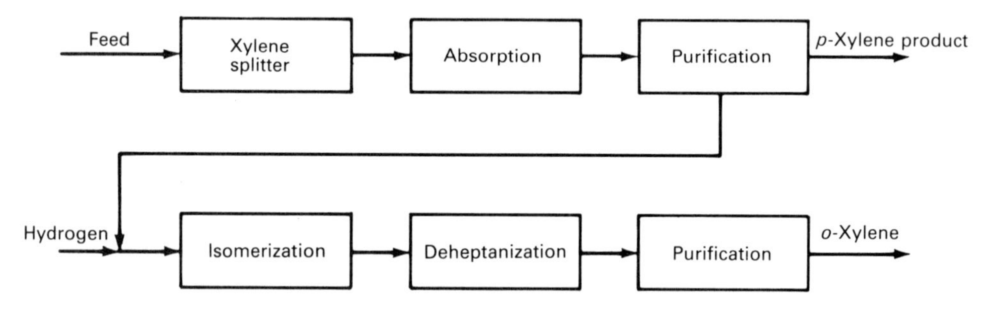

Figure 7.1 Production of xylene—UOP Parex/Isomar process

The aromatics are normally separated from the remainder of the reformate by solvent extraction and then fractionated to produce the individual aromatic compounds. As can be seen from the preceeding table, C_8 aromatics acount for about 22 percent by volume of the reformate. Some properties of the C_8 aromatics are shown in the following table:

	o-xylene	*m*-xylene	*p*-xylene	Ethylbenzene
Boiling point,°C	144.4	139.1	138.3	136.2
Melting point, °C	−25	−48	+13	−95
Density, lb/gal	7.37	7.23	7.21	7.26
Structure	CH_3, CH_3	CH_3, CH_3	CH_3, CH_3	C_2H_5

The closeness of the boiling points indicates that separation of the four isomers by distillation is difficult. Orthoxylene might be taken as a bottoms product or ethylbenzene as an overhead product, but the costs would be high. Most of the ethylbenzene used in the manufacture of styrene is produced by alkylation of benzene with ethylene. The higher melting point of *p*-xylene has resulted in most of the U.S. paraxylene being produced via crystallization at −85°F and separated from the mother liquor by filtration or centrifugation. The crude material is then concentrated in stages by remelting and crystallization or a second recrystallization at about −25°F. Filtrate from the second stage is recycled to the first stage.

In all the crystallization processes, formation of eutectics with the other isomers limits *p*-xylene recovery to about 60 percent. As mentioned earlier, an isomerization unit to convert the other isomers to the para form is often included to afford flexible yields and permit adjustment to market shifts. Isomerization processes are typically the fixed bed type, often using platinum catalysis. UOP and Englehard offer these catalysts. An alternative approach to *p*-xylene recovery is the UOP Parex process, which claims 90 percent recovery with the use of a molecular sieve adsorbent.

Mitsubishi Japan Gas Chemical has developed a process that uses HF and BF_3 to extract *m*-xylene by forming a complex. The complex is broken by heat, the HF-BF_3 recycled, and the pure *m*-xylene isomerized to the para form.

It is also possible to convert toluene to benzene and xylene via disproportionation as follows:

$$2\ C_6H_5CH_3 \longrightarrow C_6H_6 + C_6H_4(CH_3)_2$$

Arco and UOP, in conjunction with Toyo Rayon of Japan, both license vapor-phase catalytic processes to accomplish this conversion.

Market

Orthoxylene is used almost exclusively for oxidation to phthalic anhydride. Exports normally account for about 30 percent of U.S. production, although this amount is highly dependent on the state of the Western European economies, import duties, the strength of the dollar, and low-priced product from the USSR. Production in 1983 was about 855 million pounds and is projected to reach close to a billion pounds by 1988. Major U.S. manufacturers of *o*-xylene and estimated capacities are shown in Table 7-1 (2). Major *o*-xylene producers in

TABLE 7-1 Major U.S. Producers of *o*-Xylene

Producer	Location	Estimated Capacity mm lb/yr	Estimated Capacity m MT/yr
Arco	Houston, Texas	240	109
Exxon	Baytown, Texas	270	122
Conoco (DuPont)	Chocolate Bayou, Texas	25	11
Phillips	Guayama, Puerto Rico	130	59
Shell	Deer Park, Texas	125	57
Koch	Corpus Christi, Texas	110	50
Tenneco	Chalmette, Louisiana	155	70
	TOTAL	1055	478

Western Europe are shown in Table 7-2 (5), and major xylene producers in Japan are shown in Table 7-3.

The market for phthalic anhydride, and therefore orthoxylene, depends almost exclusively on the demand for flexible polyvinyl chloride, which in the early eighties was in the doldrums because of the downturn in the automobile and construction industries but recovered nicely in 1983 and 1984.

Paraxylene is used primarily in the production of terephthalic acid and dimethylterephthalate, which are intermediates in the production of polyesters. After a slow period in the first 2 years of the eighties, demand for polyester fibers has improved considerably in 1983 and 1984 because of stable petroleum prices and increasing prices for cotton. Demand in 1983 in the United States was about 3.2 billion pounds, and projections are for a level of over 3.6 billion pounds by 1987. U.S. producers are facing increasing competition from foreign

TABLE 7-2 Major Western European Producers of *o*-Xylene

Producer	Location	Estimated Capacity mm lb/yr	Estimated Capacity m MT/yr
Total Chimie	Gonfreville-L'Orcher, France	176	80
Deutsche Shell	Godorf, W. Germany	221	100
Deutsche Texaco	Heide, W. Germany	33	15
Ruhr Oel	Gelsenkirchen, W. Germany	254	115
URBK	Wesseling, W. Germany	71	32
Anic	Porto-Torres, Italy	66	30
Mobil Oil	Napoli, Italy	143	65
Montedison	Priolo, Italy	165	75
Saras Chimica	Sarroch, Italy	165	75
Esso Chimie	Rotterdam, Netherlands	176	80
PETROGAL	Leca-de-Palmeira, Portugal	66	30
CEPSA	Algeciras, Spain	88	40
	TOTAL	1624	737

TABLE 7-3 Major Japanese Producers of Xylenes

Producer	Location	Estimated Capacity	
		mm lb/yr	m MT/yr
Idemitsu Petrochemical	Tokuyama, Chiba	620	281
Maruzen Oil	Matsuyama	110	50
Maruzen Petrochemical	Chiba	150	68
Mitsubishi Chemical	Mizushima	55	25
Mitsubishi Gas Chemical	Mizushima	265	120
Mitsubishi Oil	Mizushima	620	281
Mitsubishi Petrochemical	Yokkaichi	120	54
Mitsui Petrochemical	Iwakui, Chiba	180	82
Nippon Petrochemical	Kawasaki, Ukishima	635	288
Nippon Steel Chemical	Onita	90	41
Osaka Petrochemical	Sakai	110	50
Shin-Daikyowa Petrochemical	Yokkaichi	95	43
Sumitomo Chemical	Niihama, Chiba	135	61
Teijin	Ehime, Tokuyama	375	170
Tonen	Wakayama	420	190
Toray	Kawasaki	450	204
	TOTAL	4430	2008

competitors. A large portion of synthetic fibers worldwide is being produced in countries with inexpensive labor, and they will likely back integrate into terephthalic acid and dimethylterephthalate. This back integration has already happened in places such as Mexico, and material produced there will probably compete at low prices with U.S. material in Western Europe. U.S. manufacturers of *p*-xylene and their estimated capacities are shown in Table 7-4 (3). Major

TABLE 7-4 Major U.S. Producers of *p*-Xylene

Producer	Location	Estimated Capacity	
		mm lb/yr	m MT/yr
Amoco	Decatur, Alabama	1300	590
	Texas City, Texas	1100	499
Arco	Houston, Texas	390	177
Chevron[a]	Pascagoula, Mississippi	330	150
Exxon[a]	Baytown, Texas	500	227
Phillips	Guayama, Puerto Rico	470	213
St. Croix Petro.	St. Croix, Virgin Islands	600	272
Koch[a]	Corpus Christi, Texas	390	177
Tenneco	Chalmette, Louisiana	140	63
	TOTAL	5220	2368

[a] Koch, Exxon, and Chevron are planning expansions.

p-xylene producers in Western Europe (5) and Japan are shown in Tables 7-5 and 7-6, respectively.

The long-term outlook for paraxylene is dependent on the market for polyesters. A large percentage goes into polyester fibers, although the area of largest growth will be in film and in polyester bottles.

Although the metaxylene isomer is the most plentiful in the mixed isomer stream, very little of it is used as such. Amoco produces and markets isophthalic acid produced by oxidation of a metaxylene stream. The material is used at a rate of about 100 million pounds per year, primarily in the production of alkyd surface coatings and unsaturated polyester resins.

TABLE 7-5 Major Western European Producers of *p*-Xylene

Producer	Location	Estimated Capacity	
		mm lb/yr	m MT/yr
Total Chemie	Gonfreville-L'Orcher, France	176	80
Deutsche Shell	Godorf, W. Germany	221	100
Deutsche Texaco	Heide, W. Germany	33	15
Ruhr Oel	Gelsenkirchen, W. Germany	265	120
URBK	Wesseling, W. Germany	71	32
Anic	Porto-Torres, Italy	110	50
Mobil Oil	Napoli, Italy	143	65
Montedison	Priolo, Italy	165	75
Saras Chimica	Sarroch, Italy	165	75
Esso Chemie	Rotterdam, Netherlands	176	80
PETROGAL	Leca-de-Palmeira, Portugal	66	30
CEPSA	Algeciras, Spain	397	180
	TOTAL	1988	902

TABLE 7-6 Major Japanese Producers of *p*-Xylene

Producer	Location	Estimated Capacity	
		mm lb/yr	m MT/yr
Idemitsu Petrochemical	Chiba	290	132
Mitsubishi Chemical	Mizushima	80	36
Mitsubishi Gas Chemical	Mizushima	260	118
Teijin	Ehime, Tokuyama	375	170
Tonen	Wakayama	185	84
Toray	Kawasaki	440	200
	TOTAL	1630	740

Economics

Typical production costs for paraxylene recovery by crystallization from a mixed xylene stream are as follows:

Capacity: 440 mm lb/yr (200 m MT/yr)
Capital cost: BLCC, $71 mm; OSBL, $34 mm; WC, $40 mm

	¢/lb	$/MT	%
Raw materials[a]	20.7	456	66
Utilities	5.6	123	18
Operating costs	0.8	18	3
Overhead costs	4.1	90	13
Cost of production	31.2	687	100
By-product credit[b]	(3.6)	(79)	
Net cost of production	27.6	608	
Transfer price	34.8	767	

[a] Xylenes at 14¢/lb
[b] C_9 aromatics at 14¢/lb; light ends at 8¢/lb

PHTHALIC ANHYDRIDE

Phthalic anhydride (PAN) is produced from either orthoxylene or naphthalene, and virtually all of these feedstocks are converted to the phthalic anhydride. Until the late fifties virtually all of the PAN produced in the United States was based on coal-tar naphthalene. The introduction of orthoxylene feedstock was quickly adopted by the producers, and at present most of the production is based on orthoxylene. An interesting point to keep in mind is that large-scale coal processing for synthetic fuels production could eventually substantially increase the quantity of naphthalene and reverse the trend.

Technology

With the use of both feedstocks, conversion to phthalic anhydride is accomplished by a vapor-phase oxidation employing a vanadium pentoxide catalyst as follows:

$$\underset{o\text{-Xylene}}{C_6H_4(CH_3)_2} + \tfrac{3}{2}O_2 \longrightarrow C_6H_4(CO)_2O + 3H_2O$$

$$\underset{\text{Naphthalene}}{C_{10}H_8} + \tfrac{9}{2}O_2 \longrightarrow C_6H_4(CO)_2O + 2H_2O + 2CO_2$$

In a typical system a mixture of air and orthoxylene is passed over the catalyst at 400°C. Yields of PAN are approximately 85 percent. The orthoxylene feedstock has a potentially greater yield potential: 1.40 pounds of PAN per pound of orthoxylene as compared with 1.16 pounds of PAN per pound of naphthalene. A typical block flowsheet for the process is shown in Figure 7-2.

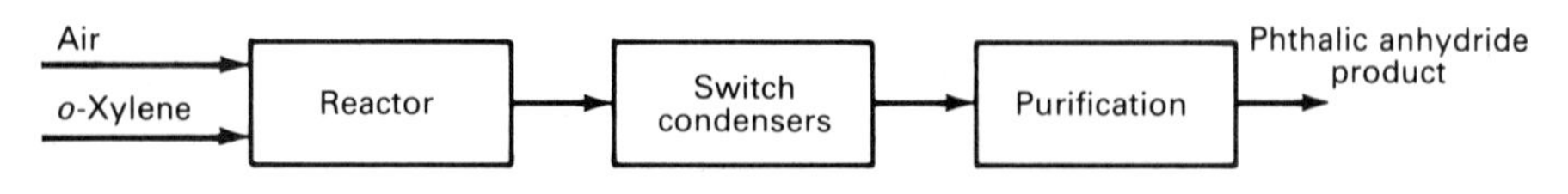

Figure 7.2 Production of phthalic anhydride

Filtered air and orthoxylene are usually preheated, the orthoxylene vaporized, and the mixture passed through a fixed bed, catalyst-filled multitube reactor or a fluid bed reactor. The exothermic reaction to phthalic anhydride takes place, and the reaction heat is often used for generation of steam. The reactor effluent is precooled and sent to a separation system, where the phthalic anhydride desublimes in switch condensers. The crude product is periodically melted and then preheated and purified by distillation. The final product can be either stored in the molten state or flaked and bagged.

When a fixed bed reactor is used, the vapor from the reactor is cooled and the phthalic anhydride is obtained in crystalline form. With a fluid bed reactor, part of the phthalic anhydride may be obtained as a liquid and part as a solid.

Market

The U.S. production of phthalic anhydride in 1983 was close to 800 million pounds, and the projection for 1987 is close to a billion pounds. Over half of the phthalic anhydride is used for the production of plasticizers for use in the manufacture of flexible polyvinyl chloride. The most common phthalate plasticizer is dioctylphthalate, which is produced by the reaction of phthalic anhydride with 2-ethylhexanol. The growth in demand for PAN, therefore, is expected to parallel the growth for flexible polyvinyl chloride.

About another quarter of the phthalic anhydride is used in the production of alkyd resins, employed almost exclusively for the production of solvent-based paints. Most of the balance is used for the production of unsaturated polyester resins. The economic recovery in 1983 and 1984 in the automotive, construction, and housing industries resulted in a good growth for the product, but the demand is not likely to reach the 1979 production levels of over a billion pounds until the late eighties. Major manufacturers of phthalic anhydride and their estimated capacities in the United States are shown in Table 7-7 (2,3). As can be seen from the table, the installed capacity is well in excess of projected demand and new installations are not expected for several years.

TABLE 7-7 Major U.S. Producers of Phthalic Anhydride

Producer	Location	Estimated Capacity		Feedstock
		mm lb/yr	m MT/yr	
Badische	Kearny, New Jersey	150	68	Purchased *o*-xylene
Exxon	Baton Rouge, Louisiana	210	95	*o*-Xylene
Koppers	Bridgeville, Pennsylvania	90	41	Naphthalene
	Cicero, Illinois	235	107	Purchased *o*-xylene
Monsanto	Texas City, Texas	165	75	Purchased *o*-xylene
	Bridgeport, New Jersey	85	39	Naphthalene
Stepan	Millsdale, Illinois	170	77	Purchased *o*-xylene
Tenn-USS	Pasadena, Texas	210	95	Purchased *o*-xylene
USS Chemicals	Neville Island, Pennsylvania	205	93	Naphthalene
	TOTAL	1520	690	

Outside of the United States installed capacity is over 2 billion pounds in Western Europe, about 800 million pounds in the Far East, and about 350 million pounds each in Eastern Europe and South America. Major producers of phthalic anhydride in Western Europe are shown in Table 7-8 (4). Major Japanese producers are shown in Table 7-9.

TABLE 7-8 Major Western European Producers of Phthalic Anhydride

Producer	Location	Estimated Capacity	
		mm lb/yr	m MT/yr
Chemie Linz	Linz, Austria	55	25
	Schwechat, Austria	77	35
SA Sopar	Zelzate, Belgium	13	6
UCB-Ftal	Oostende, Belgium	176	80
Neste Oy	Kulloo, Finland	37	17
ATOCHEM	Chauny, France	176	80
Norsolor	Villers-Saint-Paul, France	35	16
BASF	Ludwigshafen, W. Germany	243	110
Bayer	Leverkusen, W. Germany	132	60
Chemische Werke Huels	Bottrop, W. Germany	221	100
	Gelsenkirchen, W. Germany	221	100
Schelde Chemie[a]	Buttel, W. Germany	31	14
Alusuisse	Scanzorosciate, Italy	176	80
Carbochimica	Trento, Italy	75	34
Montepolimeri	Brindisi, Italy	99	45
Soc. Chim. diColleferro	Colleferro, Italy	99	45
SISAS	Pioltello, Italy	99	45
Petroles e Gas de Portugal	Lisbon, Portugal	33	15
BASF	Tarragona, Spain	35	16
CEPSA	Algeciras, Spain	44	20
	Luchana-Baracaldo, Italy	33	15
Rio Rodana	Miranda-de-Ebro, Italy	66	30

TABLE 7-8 Major Western European Producers of Phthalic Anhydride (*continued*)

Producer	Location	Estimated Capacity mm lb/yr	m MT/yr
Katalys	Nol, Sweden	33	15
Bitmac	Totton, U.K.	15	7
BP Chemicals	Hull, U.K.	165	75
	TOTAL	2389	1085

[a]The Schelde Chemie production is a by-product of anthraquinone production.

TABLE 7-9 Major Japanese Producers of Phthalic Anhydride

Producer	Estimated Capacity mm lb/yr	m MT/yr	Feedstock
Nippon Catalytic Chemical	200	91	*o*-Xylene, naphthalene
Kawasaki Chemical	150	68	*o*-Xylene
Mitsubishi Gas Chemical	140	63	*o*-Xylene
Nippon Steel Chemical	120	54	*o*-Xylene, naphthalene
Kawatetsu Chemical	65	29	Naphthalene
TOTAL	675	305	

Note: In addition, Toho Chemical has a small facility using naphthalene feedstock.

Economics

Typical production costs for phthalic anhydride via oxidation of ortho-xylene are as follows:

Capacity: 200 mm lb/yr (90 m MT/yr)
Capital cost: BLCC, $56 mm; OSBL, $16 mm; WC, $16 mm

	¢/lb	$/MT	%
Raw materials[a]	16.6	366	61
Utilities	2.8	62	10
Operating costs	1.5	33	6
Overhead costs	6.1	135	23
Cost of production	27.0	596	100
By-product credit (steam)	(3.1)	(68)	
Net cost of production	23.9	528	
Transfer price	34.7	765	

[a]Orthoxylene at 16¢/lb

TEREPHTHALIC ACID AND DIMETHYL TEREPHTHALATE

Terephthalic acid (TPA) and dimethyl terephthalate (DMT), its dimethyl ester, became available as industrial chemicals in the fifties. Together with ethylene glycol, they are the primary raw materials for the manufacture of polyesters. Since most of the material ends up as fibers, the market for these materials changes relatively rapidly as it parallels the demand for polyester fibers. Demand in the United States increased from about 2 billion pounds in 1970 to about 5.5 billion pounds in 1979 and dropped in the early eighties to slightly over 4 billion pounds as the worldwide synthetic fiber business dropped. Although U.S. producers have been hurt, Western European producers have been very hard hit and losses have been substantial. To some degree the losses in Western Europe were aggravated by imports from the United States.

Virturally all DMT or TPA is used to produce polyethylene terephthalate (PET). A still relatively small, but rapidly growing, outlet is the production of polybutylene terephthalate (PBT). This material is produced by the reaction of the DMT or TPA with 1,4-butanediol instead of with ethylene glycol.

Technology

Polyethylene terephthalate was first commercially manufactured in 1949 in England by ICI and in 1953 in the United States by DuPont. Both processes involved conversion of paraxylene to TPA by oxidation with dilute aqueous nitric acid according to the following reaction:

$$CH_3\text{-}C_6H_4\text{-}CH_3 + 4HNO_3 \longrightarrow HOOC\text{-}C_6H_4\text{-}COOH + 4NO + 4H_2O$$

The reaction is carried out in the liquid phase, and air or oxygen is introduced into the reactor. The nitric oxide is readily reoxidized to nitrogen dioxide for additional nitric acid production. Both DuPont and ICI have continuously introduced improvements in the process, but several disadvantages cannot be overcome. First, economic considerations would require a nitric acid facility to utilize the nitrogen oxides produced in the oxidation. Second, to realize adequate yields, temperature and nitric acid concentrations must be in the range where explosions are possible. Finally, the TPA produced contains impurities that make it unsuitable for most fiber applications and it requires an extensive purification process. TPA is insoluble in most common solvents, and the usual purification techniques are inadequate.

Because of the difficulty in the purification of the TPA, DMT was the initial feedstock for the manufacture of PET. If the TPA is converted to DMT, suitable purification is possible by conventional operations, such as distillation or crystallization. The conversion is as follows:

$$\text{COOH–C}_6\text{H}_4\text{–COOH} + \begin{matrix} CH_3OH \\ | \\ CH_3OH \end{matrix} \longrightarrow \text{COOCH}_3\text{–C}_6\text{H}_4\text{–COOCH}_3 + 2H_2O$$

Several companies have developed their own technologies. The major surviving technology for DMT is sometimes referred to as the Dynamit-Witten-Imhausen process and is used largely by Hercules in the United States and Hoechst in West Germany. Major contributions have been made by many companies, and the bulk of world production is based on this technology. The process involves the liquid-phase oxidation of paraxylene in acetic acid to produce DMT directly and cannot be used to produce TPA. DMT, of course, can be produced by esterification of TPA, which is actually done by many companies. The reactions are as follows:

$$\text{CH}_3\text{–C}_6\text{H}_4\text{–CH}_3 + \frac{3}{2}O_2 \longrightarrow \text{COOH–C}_6\text{H}_4\text{–CH}_3 + H_2O$$

p-Toluic acid

$$\text{COOH–C}_6\text{H}_4\text{–CH}_3 + CH_3OH \longrightarrow \text{COOCH}_3\text{–C}_6\text{H}_4\text{–CH}_3 + H_2O$$

Methyl toluate

$$\text{COOCH}_3\text{–C}_6\text{H}_4\text{–CH}_3 + \frac{3}{2}O_2 \longrightarrow \text{COOCH}_3\text{–C}_6\text{H}_4\text{–COOH} + H_2O$$

Monomethyl terephthalate

$$p\text{-}CH_3OOC\text{-}C_6H_4\text{-}COOH + CH_3OH \longrightarrow p\text{-}CH_3OOC\text{-}C_6H_4\text{-}COOCH_3 + H_2O$$

DMT

The oxidation steps usually use a cobalt acetate, naphthenate, or toluate catalyst at temperatures of about 170°C and pressures of about 215 psia. Manganese compounds have also been used as a catalyst. The esterification steps are carried out at about 150°C. The catalysts are generally recoverable and reusable. DMT can be shipped and stored in either a liquid or solid crystalline state.

Until the mid-sixties all polyethylene terephthalate was produced by DMT. However, as can be seen from the reactions, two moles of methanol are required for each mole of DMT produced. In the formation of the polymer from DMT, the methanol is reformed and, to make the process economically viable, must be recovered and recycled or sold. Therefore about 14 percent of the weight of the DMT does not end up in the polymer.

The competitive technology to the process for DMT is the Amoco process for the direct production of fiber-grade TPA. The original technology was developed in the United States by Mid-Century. The process has been widely licensed, and virtually all the worldwide production of fiber-grade TPA is based on this process. The process is a liquid-phase air oxidation, generally employing cobalt and manganese salts as a catalyst together with a source of bromine as a promoter and acetic acid as a solvent. The reactions involved are as follows:

$$p\text{-}CH_3\text{-}C_6H_4\text{-}CH_3 + O_2 \longrightarrow p\text{-}HOOCH_2C\text{-}C_6H_4\text{-}CH_3$$

p-Methyl benzyl hydroperoxide

$$p\text{-}HOOCH_2C\text{-}C_6H_4\text{-}CH_3 + O_2 \longrightarrow p\text{-}OHC\text{-}C_6H_4\text{-}CH_3 + H_2O$$

p-Tolualdehyde

$$\text{CHO}-C_6H_4-CH_3 + \tfrac{1}{2}O_2 \longrightarrow \text{COOH}-C_6H_4-CH_3$$

p-Toluic acid

$$\text{COOH}-C_6H_4-CH_3 + O_2 \longrightarrow \text{COOH}-C_6H_4-\text{CHO} + H_2O$$

p-Carboxybenzaldehyde

$$\text{COOH}-C_6H_4-\text{CHO} + \tfrac{1}{2}O_2 \longrightarrow \text{COOH}-C_6H_4-\text{COOH}$$

TPA

The major impurity is usually the *p*-carboxybenzaldehyde, which must generally be removed to form a TPA of sufficient purity for fiber application. The most common procedure is to hydrogenate the crude TPA, converting the impurity to *p*-toluic acid, which can then be converted to additional TPA. The purification to fiber-grade TPA is generally accomplished by crystallization. The technology for fiber-grade TPA also permits the production of DMT.

Although the preceding description covers the two major commercial routes to DMT or TPA, other technologies are available but are not economically attractive at this time. Various possible production schemes are shown in Figure 7-3, but most of the TPA and DMT produced worldwide at this time begin with paraxylene. Block flow diagrams for the processing schemes for DMT and TPA are shown in Figures 7-4 and 7-5, respectively.

Development work that may affect the future of DMT and TPA production is continuing. The major approaches involve either a different route to the product or a reduction in the extensive and expensive purification steps required with the Amoco process to produce fiber-grade TPA. The major effort involving an alternative route is the ammoxidation route announced by Lummus-Crest. This process is based on ammoxidation of paraxylene to produce terephthalonitrile, which is then hydrolyzed and purified to produce the fiber-grade TPA. Low yields and high residence times seem to be the prime problems. Work is proceeding on the development of new catalyst systems. In one version of the process, paraxylene and recycled *p*-tolunitrile (TN) are reacted with ammonia.

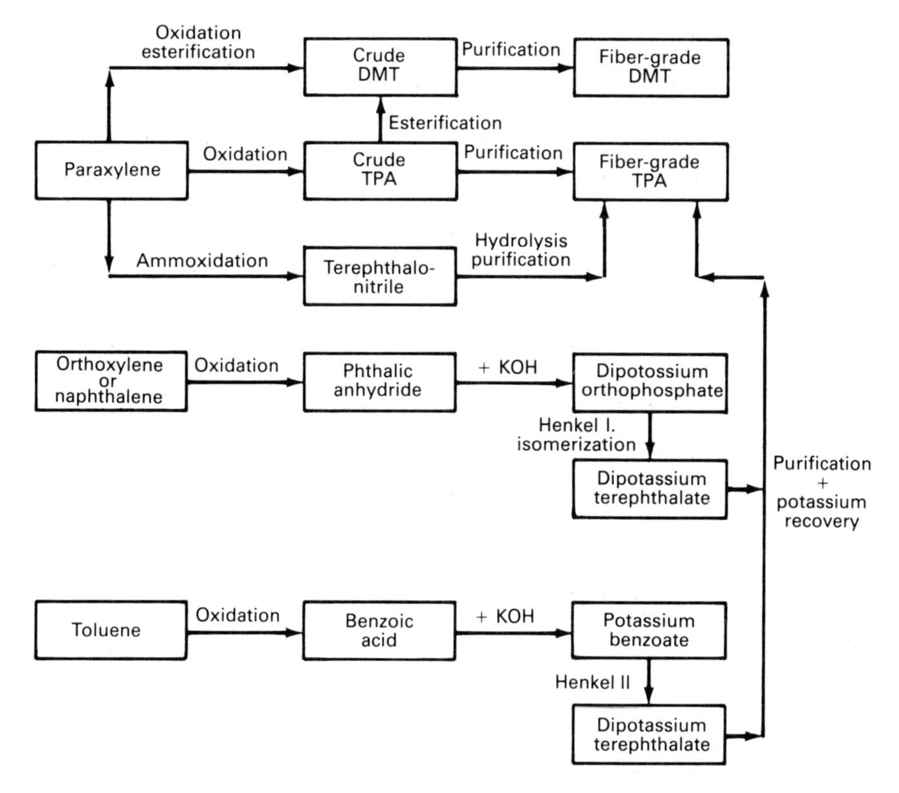

Figure 7.3 Various production schemes for DMT and TPA

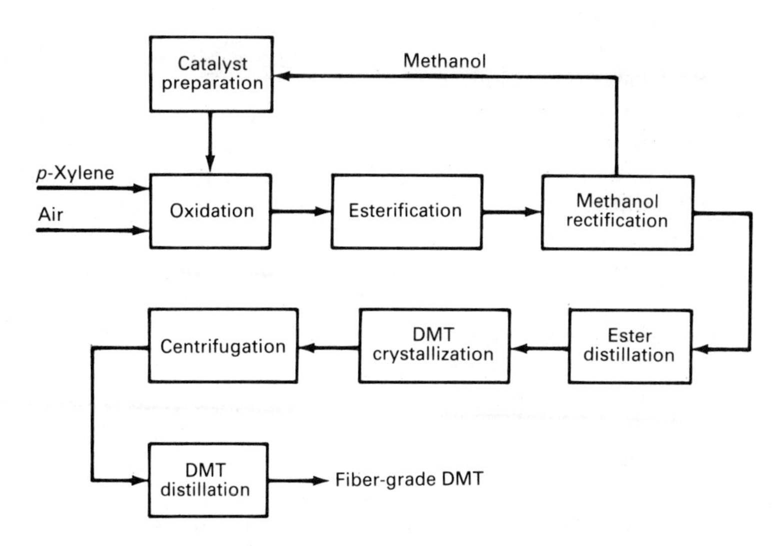

Figure 7.4 Production of fiber-grade DMT

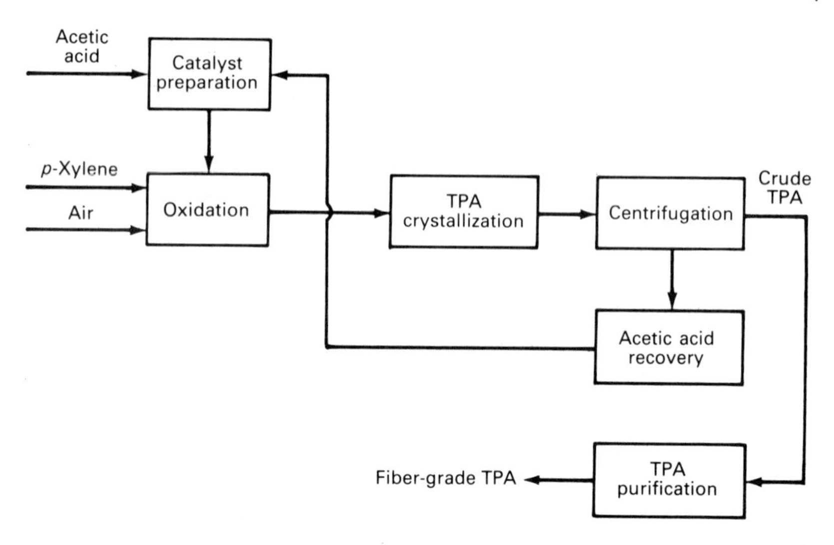

Figure 7.5 Production of fiber-grade TPA

A metal oxide catalyst is used, which, in addition to catalyzing the reaction, provides the oxygen for the reaction and is thereby reduced to a lower oxidation state. Selectivity to terephthalonitrile (TPN) and TN are claimed to be high. The TPN is hydrolyzed to produce monoammonium terephthalate (MAT). The solid MAT is then subjected to thermal decomposition to produce TPA. The TPA is then purified, possibly by crystallization.

Maruzen Oil of Japan is working on a process to minimize the extensive purification steps required in the Amoco process for fiber-grade TPA (1,2). As described previously, the oxidation of the paraxylene is accomplished in the liquid phase using acetic acid as a solvent and a catalyst system containing cobalt, manganese, and bromine. The main problem when crude TPA is used for the manufacture of fibers is the presence of *p*-carboxybenzaldehyde (PCB). The crude TPA will typically contain about 5000 ppm of the PCB, whereas the fiber-grade material will contain about 50 ppm. This impurity can act as a chain terminator during polymerization and therefore produce fibers of lower molecular weights and reduced fiber strength, and perhaps even color the final polymer. The Maruzen approach involves careful control of reaction conditions and a modified reaction system designed to assure more complete oxidation of PCB, thereby greatly simplifying the purification portion of the plant. Patent claims indicate that the reactor modifications permit reductions of the PCB to a level of approximately 200 ppm. Although this reduction is higher than the PCB level in fiber-grade TPA, Maruzen suggests that the somewhat higher value will not adversely affect the quality of the fiber produced from this material.

Market

Fibers consume a large percentage of the production of DMT and TPA, with apparel taking most of this amount. A large part of the apparel is a blend of the polyester with natural fibers, but a substantial effort is under way to improve the characteristics of the polyester yarn so that all polyester fabrics will once again be accepted by the consumer.

Polyester fibers have begun a recovery from the poor performance of 1982. In that year production in the United States dropped to 3.2 billion pounds from over 4.1 billion pounds in 1981. Capacity, as well, fell to about 4.3 billion pounds from 4.6 billion pounds. World production in 1982 decreased almost 9 percent to slightly over 11 billion pounds.

The trend worldwide is toward the use of TPA rather than DMT. The major advantages generally projected for the use of TPA are as follows. First, less raw material is required for TPA. If there is a market for the by-product methanol, this advantage is minimized. Second, a polyester facility based on TPA is generally less expensive than one based on DMT, partly because of the need to process the methanol, which is a by-product with the DMT. Exports of DMT and TPA have been steadily declining to a value of 300 million pounds in 1981. The decline is continuing as local facilities in many countries producing DMT and TPA are starting up.

Although the fibers represent the major use of the DMT and TPA, the use of polyethylene terephthalate bottle resins has been growing rapidly (8).The market among soft-drink makers and packagers has grown rapidly, and its use for liquor bottles appears likely (9). A breakdown of the end uses of DMT and TPA in the United States is as follows (12):

End Use	%
Polyethylene terephthalate fibers	82
Polyethylene terephthalate films	8
Polyethylene terephthalate resin	6
Polybutylene and other terephthalates	4

Current producers of DMT and TPA in the United States and their estimated capacities are shown in Table 7-10 (12,14). Amoco is the only company producing TPA. All the others produce DMT, although Hercofina can produce some TPA from DMT.

Western European producers of DMT and TPA are shown in Tables 7-11 and 7-12, respectively (13). Japanese producers of both terephthalic acid and dimethylterephthalate include Kuraray Yuka, Matsuyama Petrochemical, Mitsubishi Chemical, Mitsui Petrochemicals, and Toray. In addition, DMT is produced by Teijin Hercules Chemicals.

TABLE 7-10 Major U.S. Producers of DMT and TPA

Producer	Location	Estimated Capacity	
		mm lb/yr	m MT/yr
Amoco	Cooper River, S. Carolina	1000	454
	Decatur, Alabama	1530	694
DuPont	Cape Fear, N. Carolina	1250	567
	Old Hickory, Tennessee	550	249
Eastman	Columbia, S. Carolina	500	227
	Kingsport, Tennessee	500	227
Hercofina	Wilmington, N. Carolina	1300	590
	TOTAL	6630	3008

TABLE 7-11 Major Western European Producers of DMT

Producer	Location	Estimated Capacity	
		mm lb/yr	m MT/yr
Amoco	Geel, Belgium	143	65
Rhone-Poulenc	Saint-Fons, France	110	50
Dynamit Nobel	Niederkassel, W. Germany	507	230
	Steyerberg, W. Germany	331	150
Hoechst	Augsburg, W. Germany	243	110
	Offenback, W. Germany	221	100
Soc. Ital. Poliest.	Acerra, Italy	309	140
Hoechst	Vlissingen, Netherlands	187	85
Intercont. Chimica	San-Rogue, Spain	165	75
	TOTAL	2216	1005

TABLE 7-12 Major Western European Producers of TPA

Producer	Location	Estimated Capacity	
		mm lb/yr	m MT/yr
Amoco	Geel, Belgium	209	95
Rhone-Poulenc	Chalampe, France	110	50
Chimica del Tirso	Ottana, Italy	176	80
Soc. Ital. Serie Acet. Sint.	Pioltello, Italy	66	30
Intercont. Quimica	San-Roque, Spain	132	60
ICI	Wilton, U.K.	860	390
	TOTAL	1553	705

Economics

Typical production costs for TPA and DMT are as follows:

Terephthalic Acid

Capacity: 400 mm lb/yr (181 m MT/yr)
Capital cost: BLCC, $136 mm; OSBL, $59 mm; WC $43 mm

	¢/lb	$/MT	%
Raw materials[a]	17.4	384	54
Utilities	3.8	84	12
Operating costs	2.7	60	8
Overhead costs	8.6	190	26
Cost of production	32.5	718	100
Transfer price	47.1	997	

[a] Paraxylene at 22¢/lb

Dimethylterephthalate

Capacity: 400 mm lb/yr (181 m MT/yr)
Capital cost: BLCC, $118 mm; OSBL, $47 mm; WC $40 mm

	¢/lb	$/MT	%
Raw materials[a]	17.5	386	58
Utilities	3.5	77	12
Operating costs	2.5	55	8
Overhead costs	6.5	143	22
Cost of production	30.0	661	100
Transfer price	42.4	935	

[a] Paraxylene at 22¢/lb; methanol at 54¢/gal

POLYETHYLENE TEREPHTHALATE

Polyethylene terephthalate (PET) in the form of polyester fibers has dominated the synthetic fiber market with very rapid growth from the sixties until recently. However, the synthetic fiber market is quite mature, and polyester producers have been looking for new applications. In the last few years, PET has experienced explosive growth in the United States, Western Europe, and Japan for use in beverage bottles. This application, however, still represents a rather minor part of overall PET demand. Textile-grade facilities can be converted to produce high-purity PET for use in beverage bottles for a relatively modest investment.

Technology

Polyester fibers and film are the largest end users of PET, and both uses involve a low-intrinsic-viscosity polymer. About 10 percent of PET demand is accounted for by high-intrinsic-viscosity material and is used primarily to make containers such as beverage bottles.

PET is synthesized from either DMT or TPA and ethylene glycol. In either case the synthesis can be described as proceeding in two stages. The first stage results in a prepolymer—a mixture of bishydroxyethylterephthalate (bis-HET) and higher oligomers. The second stage, referred to as polycondensation, is the combining of these oligomers to form the high-molecular-weight polymer. Although the first-stage products from DMT and TPA differ slightly, the polycondensation step is essentially the same for both prepolymers. When the synthesis begins with DMT, the ester interchange reaction consists of substituting ethylene glycol for the methyl alcohol groups of the DMT and stripping the methanol, which is displaced in accordance with the following reaction:

$$C_6H_4(COOCH_3)_2 + 2HOCH_2CH_2OH \longrightarrow C_6H_4(COOCH_2CH_2OH)_2 + 2CH_3OH$$

The reaction can be catalyzed by either acidic or basic catalysts. In commercial practice zinc, manganese, and cobalt acetates are used widely. In the initial formation of the bis-HET, some dimers, trimers, and higher oligomers are always produced but do not generally present a serious problem. Probably the most important side reaction is the formation of glycol ethers from the ethylene glycol, which can affect the properties of the final polymer, usually adversely. This reaction is generally minimized by maintaining a relatively low temperature and reducing the excess ethylene glycol.

When the synthesis begins with terephthalic acid (TPA), the initial step in the polymerization sequence is an esterification, rather than a transesterification, in accordance with the following reaction:

$$C_6H_4(COOH)_2 + 2HOCH_2CH_2OH \longrightarrow C_6H_4(COOCH_2CH_2OH)_2 + 2H_2O$$

Water, rather than methanol, is liberated in the process, which simplifies the recovery section of the plant. Advantages of direct esterification include faster reaction rates and the need for less catalyst. The main difficulty involves the lower solubility of TPA in glycol and the more difficult purification of TPA (see preceding section on TPA and DMT). Commercial processes now use superatmospheric pressure and temperatures in excess of 250°C.

Whether the starting material is DMT or TPA, the second step in the polymerization sequence is the same. This step involves the polycondensation of bis-HET. When the polymer is to be used for fiber, its molecular weight is generally in the range of about 15,000 to 20,000. The reaction temperature must be above the melting point of the polymer and below the temperature at which decomposition occurs too rapidly, generally about 280°C. The removal of glycol vapors under vacuum drives the equilibrium toward polycondensation. The properties of PET are determined by the degree of polymerization, as determined by the molecular weight or the intrinsic viscosity of the resin. The commercial processes for the production of fiber-grade PET produce a polymer with an intrinsic viscosity ranging from 0.50 to 0.65. This material is usually spun directly to fiber, but some is converted into fiber-grade chips by pelletizing or band casting. The resin suitable for biaxially oriented bottles should have an intrinsic viscosity of at least 0.72 if injected molded preforms are used and 1.04 if extruded preforms are used. Although PET can be polycondensed in autoclaves or special reactors to an intrinsic viscosity of up to 0.85, variations of the polymerization process have been developed, particularly for the production of bottle-grade resins.

Block flow diagrams for the process beginning with TPA and with DMT are shown in Figures 7-6, 7-7, and 7-8. Figure 7-6 shows esterification of TPA to bis-HET; Figure 7-7 shows transesterification of DMT to bis-HET; and Figure 7-8 shows polycondensation of bis-HET, which is common to both processes.

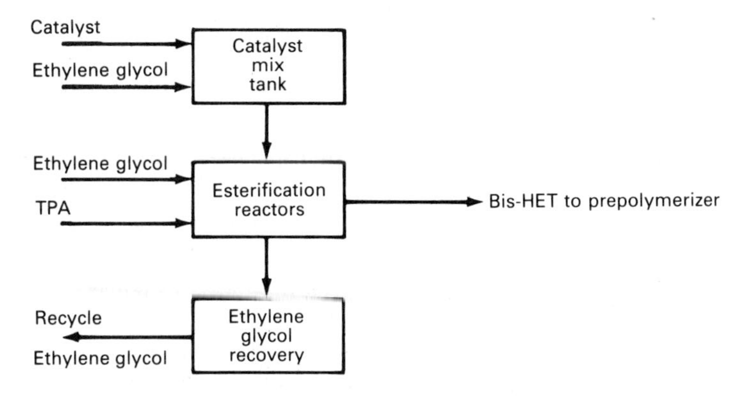

Figure 7.6 Production of bis-HET via esterification of TPA and ethylene glycol

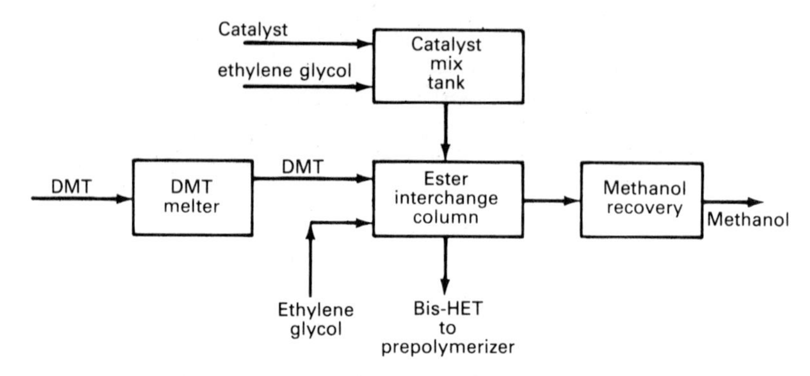

Figure 7.7 Production of bis-HET via transesterification of DMT and ethylene glycol

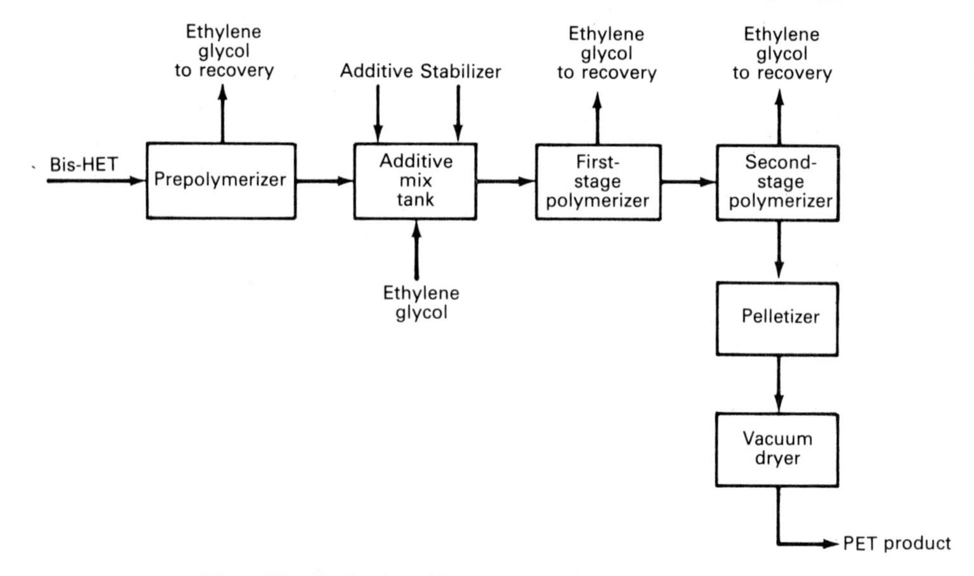

Figure 7.8 Production of PET via polycondensation of bis-HET

Market

The largest volume polyester is polyethylene terephthalate. Although polyester fibers dominate the synthetic fiber market, future growth is likely to be moderate and producers have been examining new applications. Table 7-13 lists major producers of polyethylene terephthalate in the United States (1). Tables 7-14 and 7-15 show major Western European producers of polyester fibers and polyethylene terephthalate film, respectively (2). Major Japanese producers of polyester fibers include Asahi Chemical Industry, Kanebo, Kuraray, Mitsubishi Rayon, Nippon Ester, Teijin, Toray, and Toyobo. Many of these companies also produce polyester resins.

TABLE 7-13 Major U.S. Producers of Polyethylene Terephthalate

Producer	Location	Estimated Capacity mm lb/yr	m MT/yr	Raw Material
Akzona	Lowland, Tennessee	70	32	DMT
Allied	Moncure, N. Carolina	90	41	TPA
Hoechst	Spartenburg, S. Carolina	505	229	DMT, TPA
Avtex	Lewistown, Pennsylvania	40	18	DMT, TPA
Celanese	Fayetteville, Salisbury and Shelby, N. Carolina Florence, Greenville, S. Carolina	1200	544	DMT, TPA
DuPont	Camden, Charleston, S. Carolina Kinston, Wilmington, N. Carolina Chattanooga, Old Hickory, Tennessee	1623	736	DMT, TPA
Eastman	Columbia, S. Carolina Kingsport, Tennessee	540	245	DMT
Firestone	Hopewell, Virginia	37	17	TPA
Goodyear	Scottsboro, Alabama	25	11	DMT, TPA
Tolaram Fibers	Ansonville, Asheboro, N. Carolina	32	15	
Wellman	Johnsonville, S. Carolina	50	23	
	TOTAL	4212	1911	

TABLE 7-14 Major Western European Producers of Polyester Fibers

Producer	Location	Estimated Capacity mm lb/yr	m MT/yr
Austria Faserwerke	Lenzing, Austria	31	14
Rhone-Poulenc	Bezons, France	22	10
	Gauchy and Valence, France	148	67
DuPont	Uentrop-Uber-Hamm, W. Germany	123	56
Enka	Oberbruch, W. Germany	88	40
	Obernburg-Am-Main, W. Germany	33	15
Hoechst	Augsburg, W. Germany	265	120
	Bad-Hersfeld, W. Germany	88	40
ICI	Ostringen, W. Germany	22	10
Norddeutsche Faser.	Neumunster, W. Germany	22	10
Rhodia	Freiburg, W. Germany	22	10
Spinnstofffabrik Zeh.	Berlin, W. Germany	40	18
Textilwerke Deggen.	Deggendorf, W. Germany	40	18
Lirelle	Letterkenny, Ireland	31	14
Wellman	Mullagh, Ireland	20	9
Anicfibre	Ottana, Italy	84	38
	Pisticci, Italy	35	16
	Porto-Torres, Italy	44	20
Snia Fibre	Cesano-Maderno, Italy	20	9
	Napoli, Italy	46	21
Soc. It. Poliestere	Acerra, Italy	168	76

TABLE 7-14 Major Western European Producers of Polyester Fibers (*continued*)

Producer	Location	Estimated Capacity	
		mm lb/yr	m MT/yr
Enka	Emmen, Netherlands	71	32
Finicise-Fibras Sint.	Portalegre, Portugal	53	24
Brilen	Barbastro, Spain	22	10
Nurel	Zaragoza, Spain	57	26
Polifibra	Igualada, Spain	13	6
La Seda de Barcelona	Prat-De-Llobregat, Spain	143	65
SAFA	Blanes, Spain	55	25
Ems Chemie	Ems, Switzerland	60	27
Viscosuisse	Widnau, Switzerland	88	40
Hoechst	Limavady, U.K.	18	8
ICI	Pontypool and Wilton, U.K.	115	52
	TOTAL	2087	946

TABLE 7-15 Major Western European Producers of Polyethylene Terephthalate Film

Producer	Location	Estimated Capacity	
		mm lb/yr	m MT/yr
Agfa-Gevaert	Mortsel, Belgium	49	22
Kodak-Pathe	Chalon-Sur-Saone, France	11	5
Rhone-Poulenc	St. Maurice-deBeynost, France	29	13
Agfa-Gavaert	Leverkusen, W. Germany	7	3
Hoechst	Wiesbaden, W. Germany	40	18
3M	San-Marco-Evangelists, Italy	18	8
DuPont	Contern-Hesperange, Luxembourg	66	30
ICI	Rotterdam, Netherlands	33	15
Bexford	Manningtree, U.K.	22	10
ICI	Dumfries, U.K.	55	25
	TOTAL	330	149

Economics

Approximate production costs are presented as follows for the manufacture of polyester staple from TPA and ethylene glycol and the manufacture of bottle-grade polyethylene terephthalate from both TPA and DMT.

Polyester Staple

Capacity: 100 mm lb/yr (45 m MT/yr)
Capital cost: BLCC, $79 mm; OSBL, $32 mm; WC, $30 mm

	¢/lb	$/MT	%
Raw materials[a]	60.7	1338	68
Utilities	3.9	86	4
Operating costs	7.7	168	9
Overhead costs	17.2	379	19
Cost of production	89.5	1971	100
Transfer price	122.8	2708	

[a]TPA at 47¢/lb

Bottle-Grade PET from TPA

Capacity: 60 mm lb/yr (27 m MT/yr)
Capital cost: BLCC, $23 mm; OSBL, $9 mm; WC, $14 mm

	¢/lb	$/MT	%
Raw materials[a]	52.7	1162	76
Utilities	3.3	73	5
Operating costs	3.8	84	5
Overhead costs	9.5	209	14
Cost of production	69.3	1528	100
Transfer price	84.3	1859	

[a]TPA at 47¢/lb; ethylene glycol at 31¢/lb

Bottle-Grade PET from DMT

Capacity: 60 mm lb/yr (27 m MT/yr)
Capital cost: BLCC, $23 mm; OSBL, $9 mm; WC, $14 mm

	¢/lb	$/MT	%
Raw materials[a]	54.8	1208	76
Utilities	3.7	82	5
Operating costs	3.8	84	5
Overhead costs	9.5	209	14
Cost of production	71.8	1583	100
By-product credit[b]	(2.5)	(55)	
Net cost of production	69.3	1528	
Transfer price	85.3	1881	

[a]DMT at 42¢/lb; ethylene glycol at 31¢/lb
[b]Methanol at 54¢/gal

POLYBUTYLENE TEREPHTHALATE

Initial production of polybutylene terephthalate was by Celanese in 1970, and its use as an engineering resin has grown fairly steadily. The material has several desirable properties that have accounted for its rapid acceptance. These properties include excellent electrical and mechanical characteristics, excellent resistance to chemicals and wear, and dimensional stability under load.

Technology

Polybutylene terephthalate is produced by the reaction between DMT or TPA and 1,4-butanediol. The process is essentially the same as that for polyethylene terephthalate used for the production of polyester fiber or film. Although both DMT and TPA can be used, DMT is usually preferred. In either case the synthesis is considered to occur in two stages. The first stage produces a prepolymer, which is a mixture of bishydroxybutylterephthalate and higher oligomers. The second stage, polycondensation, results in combinations of the oligomers to produce the high-molecular-weight polymers.

$$C_6H_4(COOCH_3)_2 + 2HO(CH_2)_4OH \rightleftharpoons C_6H_4(COO(CH_2)_4OH)_2 + 2CH_3OH$$

$$C_6H_4(COO(CH_2)_4OH)_2 \rightleftharpoons HO\left[(CH_2)_4OOC{-}C_6H_4{-}COO\right]_n(CH_2)_4OH + (n-1)HO(CH_2)_4OH$$

The first stage, a transesterification reaction, is affected by temperature, excess glycol, catalyst, and reaction time. Depending upon these conditions, several different oligomers can be obtained. Formation of glycol ethers is the dominant side reaction. Effective catalysts for the transesterification step include cobalt, zinc, lead, manganese, and calcium acetates. Antimony trioxide appears to be particularly effective for the polycondensation step. PBT is generally produced from DMT via batch processing. A typical block flow diagram is shown in Figure 7-9.

Batch processing permits small-volume production of thermoplastics with varying properties. One advantage of using DMT is that it can be melted and

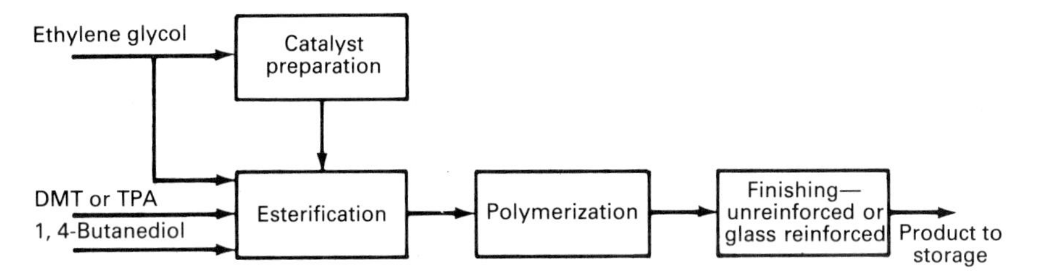

Figure 7.9 Production of PBT

fed to the reactor as a liquid. If TPA is used, it must be introduced as a solid. The esterification step produces by-product methanol in the DMT-based process. The methanol is impure. In the TPA-based route, water is produced as a by-product. The equipment for PET fiber facilities can be, and often is, converted to PBT production. Once the PBT polymer chips are available, they can be mixed with additives and pigments and, in many applications, reinforced with glass fibers. As PBT demand increases, further consideration is likely to be given to continuous processing to improve production economics. A major advantage of continuous polymerization is that direct extrusion from the melt is feasible, which would eliminate polymer chip production.

Market

The selection of an engineering thermoplastic for a particular use involves, in addition to price, an evaluation of the properties and processing costs of the finished end product. Many competing resins are available, and an optimum choice is difficult. PBT has enjoyed rapid acceptance because of the following properties:

1. Excellent electrical and mechanical properties.
2. Good resistance to wear and chemicals.
3. Low water absorption.
4. Good fatigue endurance.
5. Dimensional stability under load and in moist environments.

Major applications of PBT are in the automotive, electrical, and electronic industries. Consumption in the United States in 1981 was approximately 60 million pounds and, for the next year or two, was in the doldrums because of the worldwide recession. Although the major use of PBT in the automotive industry has been for exterior applications, future use will probably be for under-the-hood electrical systems and perhaps even body panels. Consumption in 1984 was about 80 million pounds per year and is projected to reach 130 million pounds per year by 1990 (6).

Table 7-16 indicates major U.S. producers of PBT and their estimated capacities (2). Major Western European producers are shown in Table 7-17 (estimated capacities are not available) (3). European consumption in 1984 was about 50 million pounds per year, and this amount is expected to grow at 10 percent per year through 1990 (6).

TABLE 7-16 Major U.S. Producers of PBT

Producer	Location	Estimated Capacity	
		mm lb/yr	m MT/yr
Celanese	Bishop, Texas	70	32
GAF	Calvert City, Kentucky	10	5
General Electric	Mt. Vernon, Indiana	130	59
	TOTAL	210	96

TABLE 7-17 Major Western European Producers of PBT

Producer	Location
ATOCHEM	Serquigny, France
BASF	Ludwigshafen, W. Germany
Ciba-Geigy	Lampertheim, W. Germany
Montepolimeri	Ferrara, Italy
Akzo Plastics	Emmen, Netherlands

Economics

Typical production costs for polybutylene terephthalate from DMT are as follows:

Capacity: 22 mm lb/yr (10 m MT/yr)
Capital cost: BLCC, \$10 mm; OSBL, \$4 mm; WC, \$8 mm

	¢/lb	\$/MT	%
Raw materials[a]	88.0	1940	79
Utilities	3.9	86	4
Operating costs	5.0	110	5
Overhead costs	14.2	313	12
Cost of production	111.1	2449	100
Transfer price	130.2	2871	

[a] DMT at 42¢/lb; 1,4-butanediol at 80¢/lb

REFERENCES

Xylenes

1. List, H. L. "Xylenes—A Status Report," *International Petrochemical Developments,* Vol. 2, No. 24, December 15, 1981.
2. "Chemical Profile," *Chemical Marketing Reporter,* August 15, 1983.
3. "Chemical Profile," *Chemical Marketing Reporter,* August 22, 1983.
4. "Petrochemical Handbook '83," *Hydrocarbon Processing,* November 1983.
5. *1984 Directory of Chemical Producers—Western Europe,* SRI International.
6. *Kirk-Othmer Encyclopedia of Chemical Technology,* Third Edition, Vol. 24. New York: Wiley, 1984, pp. 709–44.
7. *JCW Chemicals Guide 82/83.* Japan: The Chemical Daily Co. Ltd., March 1982.
8. *Chemical Industry Yearbook,* Second Edition. Surrey, England: Industrial Press, 1984.

Phthalic Anhydride

1. List, H. L. "Phthalic Anhydride—A Status Report," *International Petrochemical Developments,* Vol. 2, No. 23, December 1, 1981.
2. "Chemical Profile," *Chemical Marketing Reporter,* July 1, 1983.
3. *1984 Directory of Chemical Producers—United States,* SRI International.
4. *1984 Directory of Chemical Producers—Western Europe,* SRI International.
5. "Petrochemical Handbook '83," *Hydrocarbon Processing,* November 1983.
6. *Kirk-Othmer Encyclopedia of Chemical Technology,* Third Edition, Vol. 17. New York: Wiley, 1984, pp. 732-77.
7. *JCW Chemicals Guide 82/83.* Japan: The Chemical Daily Co. Ltd., March 1982.
8. Verde, L. and M. Neri. "Make Phthalic Anhydride with Low Air Ratio Process," *Hydrocarbon Processing,* November 1984, pp. 83–85.
9. *Chemical Industry Yearbook,* Second Edition. Surrey, England: Industrial Press, 1984.

TPA and DMT

1. U.S. Patent 3,839,437 (October 1, 1974) to Maruzen Oil.
2. U.S. Patent 3,846,487 (November 5, 1974) to Maruzen Oil.
3. List, H. L. "TPA and DMT—A Status Report," *International Petrochemical Developments,* Vol. 1, No. 3, December 15, 1980.
4. "Bid for a Better Polyester Image," *New York Times,* April 4, 1983.
5. "Major Synthetic Fibers Well Into Long Awaited Recovery," *Chemical and Engineering News,* May 30, 1983.
6. "Man Made Fibers Production Sagged Worldwide in 1983," *Chemical and Engineering News,* February 28, 1983.
7. "Fiber Raw Materials Follow Fibers Downturn," *Chemical and Engineering News,* July 5, 1982.
8. "Chemical Profile," *Chemical Marketing Reporter,* April 27, 1981.
9. "PET Grabs for the Liquor Bottle," *Chemical Week,* November 24, 1982.
10. "Chemical Profile," *Chemical Marketing Reporter,* April 5, 1982.
11. "Petrochemical Handbook '83," *Hydrocarbon Processing,* November 1983.

12. "Chemical Profile," *Chemical Marketing Reporter,* August 8, 1983.
13. *1984 Directory of Chemical Producers—Western Europe,* SRI International.
14. *1984 Directory of Chemical Producers—United States,* SRI International.
15. *Kirk-Othmer Encyclopedia of Chemical Technology,* Third Edition, Vol. 17. New York: Wiley, 1984, pp. 732–77.
16. *JCW Chemical Guide 82/83.* Japan: The Chemical Daily Co. Ltd., March 1982.
17. "Key Chemicals DMT/PTA," *Chemical and Engineering News,* December 17, 1984, p. 14.

Polyethylene Terephthalate

1. *1984 Directory of Chemical Producers—United States,* SRI International.
2. *1984 Directory of Chemical Producers—Western Europe,* SRI nternational.
3. "Petrochemical Handbook '83," *Hydrocarbon Processing,* November 1983.
4. *JCW Chemicals Guide 82/83.* Japan: The Chemical Daily Co. Ltd., March 1982.
5. "Materials '85," *Modern Plastics,* January 1985, pp. 61–71.
6. "Outlook Unsettled After Boom in Fibres," *European* Chemical News, February 18, 1985, p. 10.

Polybutylene Terephthalate

1. List, H. L. "Polybutylene Terephthalate—A Status Report," *International Petrochemical Developments,* Vol. 3, No. 8, April 15, 1982.
2. *1984 Directory of Chemical Producers—United States,* SRI International.
3. *1984 Directory of Chemical Producers—Western Europe,* SRI International.
4. *Kirk-Othmer Encyclopedia of Chemical Technology,* Third Edition, Vol. 18. New York: Wiley, 1984, pp. 549–74.
5. "Materials '85," *Modern Plastics,* January 1985, pp. 61–71.
6. "Engineering Resin Makers Scramble to Establish Position As Mobay Launches New Line," *Chemical Marketing Reporter,* February 11, 1985.

Appendix A: Basis for Production Cost Economics

The approximate production costs presented in the book are based on operations in the Gulf Coast of the United States under conditions projected to exist at the beginning of 1986. Capital costs are based on a *CE* Plant Cost Index of 330. Adjustments can be made by multiplying the cost based on 330 by the ratio of the current index, as presented weekly in *Chemical Engineering* magazine, to 330.[1] Unless specific information is available, other assumptions are as follows:

1. Working capital is equivalent to 4 months of production costs.
2. Depreciation is taken as straight line with a 10-year life and no salvage value.
3. Maintenance is taken as 6 percent of battery limits capital cost.
4. Direct overhead is taken as 40 percent of labor and supervision.
5. General plant overhead is taken as 65 percent of operating costs.
6. Insurance and property taxes are taken as 1.5 percent of total fixed investment.
7. Return on investment is taken as 30 percent of total fixed investment before taxes.
8. Interest on working capital is taken at 10 percent annually.

The major contributors to the production costs for virtually all petrochemicals are the raw material costs, usually followed by overhead costs, in which the major contributor is the capital cost investment. In each case the price of the major raw material used has been indicated. With this information and the CE cost index described above, approximate adjustments can be made to the major contributors to the production costs to get approximate figures for variations in time and capital costs.

[1] For the basis of this index, refer to "CE Plant Cost Index—Revised," *Chemical Engineering*, April 19, 1982, p. 153.

INDEX